VISIONS
★
APPARITIONS
★
ALIEN VISITORS

Key to front cover illustrations:

(1)	(2)	(3)
(4)	(5)	(6)

(1) Spirit of the dead raised at Walton-le-Dale; from *The Necromancer of the Nineteenth Century.*
(2) Fairies, dragon flies and other strange creatures by Warwick Goble; from *The Book of Fairy Poetry* (1920).
(3) Guardian Angel; a German postcard.
(4) Fairy, by Warwick Goble; illustration for Charles Kingsley's *The Water Babies.*
(5) Humanoid seen by M. Masse at Valensole, France (1965).
(6) Men in Black, from Hilary Evans, *UFOs: The Greatest Mystery* (reproduced by permission of Rainbird/Albany).
All illustrations supplied by The Mary Evans Picture Library.

VISIONS
★
APPARITIONS
★
ALIEN VISITORS

by

Hilary Evans

THE AQUARIAN PRESS
Wellingborough, Northamptonshire

First published 1984

British Library Cataloguing in Publication Data

Evans, Hilary
    Visions ★ Apparitions ★ Alien Visitors.
    1. Psychical research
    I. Title
    133.8   BF1031

    ISBN 0-85030-414-8

The Aquarian Press is part of the Thorsons Publishing Group

Printed in Great Britain by
Butler & Tanner Ltd, Frome and London

CONTENTS

LIST OF ILLUSTRATIONS

8 VISIONS ★ APPARITIONS ★ ALIEN VISITORS

14b. Photograph of the materialization of Queen Astrid during a seance at Kobenhavn in 1938. From Gerloff, *Die Phantome von Køpenhagen*.
15. A succubus as depicted by Robin Ray.
16. The visionaries of Garabandal in ecstasy before a vision of the Virgin, 1961. From F. Sanchez-Ventura y Pascual, *Las apariciones no son mito* (1966).

ACKNOWLEDGEMENTS

The length of my bibliography indicates my indebtedness to others who have gone before me, and at the same time reveals how widely scattered has been the material that has been brought together in this study of the experiences of peasant girls and theologians, mountaineers and circumnavigators, flying saucer cultists and demonologists, lonely children and eminent authors, kindly motorists and frightened soldiers — and a countless number of very ordinary people. My debt to researchers who have studied specific aspects of the entity problem is of course very great, and without those who had the courage to report their experiences in the first place, this experienced-based study could not exist. I hope I can encourage the reader to go back to the original sources, for no second-hand sampling can do justice to the weight of evidence contained, for example, in the publications of the Society for Psychical Research over the past hundred years.

Under these circumstances individual acknowledgement is not only invidious but impossible; so I trust that all those cited will accept my thanks. However, I do feel a special obligation to three researchers who have not only dared to present very radical findings in public, but have also shown themselves willing to face up to their implications. These are Morton Schatzman, whose *The Story of Ruth* is one of the most revealing case studies ever published; David Hufford, whose *The Terror That Comes in the Night* is a model study not only for its research-based approach and broad perspective, but also for its courageous acceptance of findings that run against accepted psychological notions; and Alvin Lawson, who performed practical experiments where others had been content with armchair speculation, and thereby gave us scientifically testable hypotheses in a field which had hitherto been regarded as entirely anecdotal.

To these, and to all others who are prepared to face the challenge of the anomalous and go where the evidence leads them, I offer my thanks and my respect.

Was Luther's picture of the Devil less a reality, whether it were formed within the bodily eye, or without it?

CARLYLE, *Sartor Resartus (1833)*

And if fairies actually exist as invisible beings or intelligences, they are natural and not supernatural, for nothing which exists can be supernatural; and therefore it is our duty to examine them just as we examine any fact in the visible realm wherein we now live, whether it be a fact of chemistry, of physics, or of biology.

W. Y. EVANS WENTZ, *The Fairy-Faith in Celtic Countries (1911)*

The subject of apparitions has for centuries occupied the attention of the learned, but seldom without reference to superstitious speculations. It is time, however, that these illusions should be viewed in a perfectly different light; for, if the conclusions to which I have arrived be correct, they are calculated, more than almost every other class of mental phenomena, to throw considerable light upon certain important laws connected with the physiology of the human mind.

SAMUEL HIBBERT, *Sketches of the Philosophy of Apparitions (1825)*

PREFACE

Many different kinds of people have many different kinds of experience in which they seem to see a more-or-less human-like figure which there are good reasons for believing is not as 'real' as it seems to be. But if it is not real in the accepted sense, is it real in any other sense, or totally unreal? And are all these varied experiences real/unreal in the same way and to the same degree? Are they, indeed, the same kind of experience in different forms, or a variety of experiences with a superficial similarity?

To the best of my knowledge, no such field survey has been undertaken previously. No doubt one of the reasons is the sheer magnitude of the task: students of these phenomena have tended to stick to a single category, feeling they had quite enough on their hands with ghosts, say, without also taking in visions of the Virgin Mary. But another reason seems to have been a failure to appreciate the relevance of such comparisons: an eminent writer on ghosts, for example, was astonished that I could be bothered with religious visions. If that particular point of comparison is not self-evident, how much less so must be the juxtaposition of 'childhood playmates', magically conjured spirits, extraterrestrial visitors and *doppelgängers*, to take just a few of the picturesque assortment we shall encounter in the course of our inquiry.

The quantity of material available is indeed colossal. To take just one example: between 1928 and 1975, 230 cases of visions of the Virgin Mary were reported with sufficient conviction to justify the attention, if not the acceptance, of the Catholic Church.[21] A lifetime would hardly suffice to familiarize oneself with every relevant instance, and new cases are being added almost daily. I ask the reader to bear in mind, therefore, that a study such as this has to be very selective, and that one person's selection might not have been another's. Even if I knew of every species and sub-species of entity sighting, I would not have been able to include them — simply to list the entities reported by anthropologists and folklorists would result in a vast catalogue, not to speak of discussion and analysis. What I do feel justified in claiming is that this study examines a sufficiently representative selection to warrant forming some working hypotheses that are unlikely to be seriously invalidated by future findings.

Formulating hypotheses and drawing conclusions, though they are the natural culmination of the inquiry, are nevertheless something of a postscript to the main study. Most of it consists of an examination of the material, to see what each category demands in the way of explanation: I hope this will be seen as valuable even if the reader eventually draws from it quite different conclusions than I do.

The plan I have followed is this:

First, an introductory section, which treats of general matters such as what we mean by 'entity', 'identity' and 'reality'.

Part One: *The Entity Experience.* Various forms in which entities manifest spontaneously to percipients.

Part Two: *Experimental Entities.* Various ways in which people have deliberately tried to cause entities to manifest.

Part Three: *Explaining the Entity Experience.* Relevant studies which throw light on our study, and some hypotheses as to how these phenomena may be accounted for.

A note on evidence

In the course of this study I shall be citing a great many case histories by way of example. Let us face the fact here and now that virtually every one of these cases is purely anecdotal, comprising witness testimony to an event for which that testimony generally constitutes the only evidence. It is not unlikely that one or more of these cases is in one way or another spurious: I have reported them as they were reported by the percipient, but I would not care to say of any single case that it is beyond question genuine. It is indeed possible that every one of the witnesses I quote is either lying or deluded, even in those cases where more than one person, strangers to one another, report the same experience; it is also possible that in all the many thousands of other cases which I could have cited in place of those I chose, the witnesses were either lying or deluded. If you find that easier to believe than that these persons really did have these experiences, I recommend you to stop reading here and now, for there is really no point in your proceeding unless you are willing to accept that the majority of these people really did, *in some form*, have the experience they reported.

What form that experience seemed to take, and what its real form was, is another matter — indeed it is the matter of this book. In those cases where the witnesses offered an interpretation of their experience, we shall not necessarily accept that interpretation. But in a very great many cases they did not do so; they simply said what had happened to them, and hoped that someone else would explain it.

A note on terminology

Words and phrases exist to meet most of the requirements of this study, but not all; and sometimes the terms can be used in more than one connotation. I hope I can avoid confusion by the following usages:

The person who 'sees' an entity is the *percipient* — who could be you, me or anyone.

To avoid tiresome circumlocution, I use the word *see* to signify the act of perception, though it is in many cases far from clear to what extent, if any, the organs of sight are involved.

The *entity* is the more-or-less human-like figure the percipient thinks he sees. It could be the spirit of Aunt Agnes materializing from the dead at a seance; it could be a vision of the Virgin Mary; it could be a spaceperson from an Unidentified Flying Object, and so on.

The *apparent* is the person the entity appears to be — the real Aunt Agnes, the Virgin Mary in person, an actual Ashtar from Alpha Centauri.

The *agent* is the person, force or whatever who is responsible for the entity and for enabling or causing it to be seen by the percipient. For example, if the apparent is your Aunt Agnes, it may be the real Aunt Agnes who is the agent; or it may be your own imagination; or it may be Satan seeking to deceive you; or it may be a mischievous elemental playing games with you.

To avoid irritating repetition I shall not always use the word *alleged* before each report. But please bear in mind that a very great number of the experiences we shall be considering depend upon the unsupported testimony of a single individual, and however ready we may be to believe him, we cannot accord his statement more than a provisional status.

The meaning we give to the word *real* is so central to our inquiry that we shall have to discuss it more fully before we embark on our study. By and large, though, I use it to denote whatever appertains to our seemingly objective universe. It implies a capability of being detected by the normal senses of any healthy individual who is suitably located, and by appropriate instruments such as a camera.

If only it could be that simple.

INTRODUCTION
THE ENTITY ENIGMA

One day, when she was twelve years old, Glenda came home from school and went upstairs to her room in the Dagenham council house where she lived with her parents and sister. A little while later she realized she was not alone in the room; with her was a stranger, female, whom she would later describe as a 'spacewoman' and who, in the drawing Glenda made of her, looks like a character from a television science fiction series. This entity was later to appear intermittently in Glenda's life — sometimes when awake, sometimes in dreams; neither malevolent nor benevolent, though seemingly concerned for Glenda's well-being. An elusive, enigmatic apparition.

By any objective means of assessment, Glenda's spacewoman does not exist. For centuries, the great majority of scholars and scientists have not felt it necessary to provide any more precise explanation for such reports than that they are figments of the percipient's imagination. There are two reasons why we should feel dissatisfied with dismissing Glenda's experience in such a way. First, because such a non-explanation is no help to Glenda at all when she asks us what it was that she experienced: and it is disgraceful that when a human being asks for help, she should be refused it. And second, because in neglecting the investigation of such reports, science is passing over a wealth of case-material which could tell us more about how the human mind works than any amount of behaviourist research based on the response-process of rats in mazes or the learning abilities of apes.

If experiences like Glenda's were rare, there might be some excuse for not devoting time and resources to studying them. But they are not rare. Glenda is one of countless percipients who have reported seeing figures, more or less human in appearance, which they know are not real, in the sense of being objectively present as a friend or colleague might be present, but which have a kind of reality — often overwhelming in its impact — for those who see them. Glenda's experience is one of a category sufficiently frequent to have earned a label — the 'bedroom invader': but that is a label only, not a definition and still less an explanation. Other entities which have been seen have been interpreted as ghosts, religious and mystical visions, spirit manifestations of the dead, hallucinations, OBE (out of the body) projections, *doppelgängers*, thought-forms, folklore figures, impersonating demons, guardian angels and extraterrestrial UFO occupants. But these, too, are only labels.

Such entities have been reported throughout history, frequently and widely enough to qualify as a scientific anomaly far more prevalent than many of the things that scholars and scientists have studied; but their appearance is so arbitrary, so intermittent, so elusive and so subjective as to give scientists ground for refusing to acknowledge them as suitable subjects for investigation. They have been able to justify their refusal on a variety of specious grounds, most frequently that the phenomena do not belong to their particular field. The fact that visionary experiences are so often associated with religious faith is a convenient let-out — or rather was, when there was more religious faith about. This is not an area in which most scientists feel happy — they have tended to disbelieve, or to regard their personal beliefs and their scientific activities as two separate things sharing no interface. Only the rare scientist recognizes the challenge and accepts it.

It is indeed far from clear which disciplines are best qualified to probe these phenomena, and given the specialist nature of most scientific work it is probable that only a scientist whose capability ranged over a number of disciplines would be able to approach them with a sufficiently open mind. Important contributions have been made by behavioural scientists, though it is noteworthy that the anthropologist and the sociologist have on the whole contributed more than the psychologist. Yet more often than not the most helpful contributions have been made by totally unqualified amateurs, like the Jesuit priest Father Herbert Thurston, who devoted much of his lifetime's work to researching these subjects in an admirably scientific spirit.

Beneath the superficial reasons for the scientists' reluctance lies a more fundamental explanation, which is clear enough to those of us who are fortunate enough not to be professional scientists. For the fact is that these phenomena, once admitted as fact, will necessitate a revision of the existing structure of scientific knowledge, no matter what they turn out to be. Only if they can be summarily dismissed as figments of the imagination, hallucinations, the meaningless products of sick minds, can they be accommodated within our existing belief-system without our having to make any major adjustments.

But can they be summarily dismissed in such a way? That, in large part, is what this study is about.

The confusion of diversity

Entity experiences vary as to *what* is seen and *how* it is seen, as well as when, where, by whom, and under what conditions and circumstances. Each of these variables further complicates and confuses a range of experiences which is already sufficiently diverse to confuse us all. For this reason, such studies as have been made of these phenomena have tended to concentrate on a single category — on ghosts, or religious visions, or UFO-entities. This implies two assumptions: first, that the category is homogeneous ('all ghosts have something fundamental in common') and second, that it is distinct from other categories ('what all ghosts have in common is different from what all religious visions have in common'). This, of course, implies that we know enough about ghosts and visions to say when an entity experience is one or the other.

If we knew that ghosts and visions were two distinct things, this in itself would be a valuable step forward, for to make that distinction we would have had to have acquired some definite information as to their nature. Perhaps we will eventually find that there *are* these essential differences, but here and now we have no right to suppose so: given our current knowledge, any study confined to a single category is premature and dangerous. For if the differences between the supposed categories should turn out to be more apparent than real, relying on them could lead to false conclusions. At this stage of our ignorance, a comparative study is both the scientific approach and the one most likely to be informative. And certainly the single-category studies of the past, though they have helped us to ask more and more pertinent questions, have yet to lead to acceptable answers.

During the primitive phases of man's cultural history, entity experiences were regarded as more or less religious in nature; the entities themselves were interpreted as being either divine or demonic, and priests and shamans, magicians and sorcerers, according to the prevailing ethos, were regarded as the appropriate authorities to deal with them.

A more open debate came with the Renaissance in western Europe, and in the sixteenth and seventeenth centuries some very perceptive works were published that can still be studied with advantage — for example, the sixteenth-century treatise *Of Ghosts and Spirits Walking by Night*. [87] The growth of institutionalized science, however, for all its massive benefits in its chosen fields, had the effect of inhibiting the study of phenomena that occurred outside those fields: the very existence of such phenomena as ghosts and visions had to be denied, to preserve the coherence of the splendid belief-systems which the scientists were so industriously assembling.

During the nineteenth century the pendulum, to a small degree, swung back again. A number of people in various walks of life, but most notably and significantly writers and doctors, protested against the official embargo, and fierce controversies were sparked off first by mesmerism and then by spiritualism. A few scientists felt it was their duty to take anomalous phenomena seriously, but even after the passage of another hundred years they remain few and far between. Today the world is still in very much the situation it has been in for most of its history, in which a great many people are having extraordinary experiences which the intellectual Establishment — which today means the scientific Establishment — refuses to acknowledge as a legitimate field for research. As a consequence, there is nobody to whom Glenda can turn for an answer to her question, 'What was it that I saw?' Her

doctor, her priest, her psychoanalyst, the police, the press — all might
conceivably show a sympathetic interest; but they are unlikely to be
able to provide an acceptable explanation, and even if they do, it is
unlikely that all will agree on the same explanation. Glenda's entity
is likely to remain an enigma so far as the authorities are concerned.

★   ★   ★

Because the subject of entity-experiences has never been tackled as a
whole, there is little awareness of how widespread this phenomenon
is. Indeed, no reliable figures are available for any of the categories,
and for some there are no quantitative indications whatsoever.
However, there are one or two straws in the wind. Thus, in 1893-4
the SPR (Society for Psychical Research, London) conducted a Census
of Hallucinations[148] which established that, of the more than 15,000
people questioned, approximately one in ten had at one time or another
experienced a hallucination of some kind. Since there is nothing to
suggest that entity-experiences are particularly prevalent among any
category of percipients rather than another — whether by age,
geographical location, ethnic origin or socio-economic classification
— there is little risk if we project this figure for the population as a whole.
That is to say, in a country with a population of sixty million, some
six million persons believed themselves to have had an entity-experience
of the apparition or hallucination type. Not many fields of scientific
study can point to six million case histories!

A statistic of a different kind is provided by recent studies of visions
of the Virgin Mary. Gilbert Cornu, a distinguished French ufologist,
has established that there were 230 cases of alleged visions of the Virgin
Mary between 1928 and 1975.[21] Some of these were witnessed by
literally hundreds of percipients. While the incidents at Lourdes, Fatima
and Garabandal, and to a lesser extent La Salette and Beauraing, have
been widely publicized, the astonishing number of such reports is not
generally known. Moreover, Cornu's figure relates only to those cases
reported to, and officially acknowledged by, the Catholic Church
authorities. (It must be pointed out that this acknowledgement implies
no endorsement: it really amounts to no more than that the Church
is not prepared to dismiss the report out of hand. As we shall see, the
Church officially takes a hard-headed attitude in these matters.)

Entities of a seemingly different sort have been widely reported in
connection with Unidentified Flying Objects, themselves, of course,
hardly recognized as a legitimate field for scientific study. An indication
of the extent of the UFO-entity phenomenon is given by David Webb's
report for CUFOS on the 1973 sightings[164]: he logged seventy reports
of humanoid entities during a five-month period from August to
December 1973. If this figure were typical, and if humanoid entities
have been seen at a steady rate since the outbreak of the modern UFO

INTRODUCTION 19

wave in 1947, this would mean that some 5700 sightings have occurred between 1947 and today. Of course, the 1973 frequency may *not* have been typical, and the rate of experiences may *not* have been steady; on the other hand, it is also true that there must have been many cases which Webb did not learn about, not to mention the fact that, with only a handful of exceptions, his report is confined to the United States.

Other such catalogues have been compiled, such as Basterfield's collection of UFO-entities reported in Australasia, which listed 67 such cases[5]. While these samplings are too scattered to present a firm statistical base, they are sufficient to indicate that a great many people, from different cultures and from a wide geographical spread, believe themselves to be seeing and even encountering extraterrestrial entities.

Apart from these attempts at methodical study, there are innumerable collections of cases gathered in anecdotal form over the centuries. We have such collections dating from the seventeenth century, many from the nineteenth, and a veritable flood from the twentieth. Granted that the scientific value of this anecdotal material is negligible, the fact that such reports continue to be made, on so wide a scale, in defiance of Establishment opinion, makes it sufficiently clear that to have an entity experience — whatever that implies — is no very uncommon thing. Though you yourself may not have had such an experience, there is a very good chance that someone in your family, or your neighbour, or a colleague at work, has had one. It is also likely that they have not told anyone about it. Time and time again investigators, while researching one incident, have stumbled accidentally on others that were never revealed outside the percipient's immediate acquaintance.

Why this should be so, is plain enough. Fear of ridicule at the least, fear of being considered mentally unstable or even ill, are sufficient deterrents to stifle, in perhaps a majority of witnesses, their natural desire to share their experience with others and obtain an explanation, or reassurance, or the comfort of knowing that others have had similar experiences.

There is still a strong taboo in operation that covers all aspects of the paranormal, muffling both the witness who undergoes the experience and the investigator who seeks to find an explanation for it: once it was the established Church that imposed the stigma, today it is official science. That is bad enough, but what is perhaps worse is that this attitude is then adopted by the public at large. Many UFO witnesses have been viciously harassed in the United States, even in this enlightened age and in a country exposed to such a variety of shades of opinion. By refusing to recognize entity-experiences as a legitimate area for serious study, by allowing witnesses to believe that their experience must be the consequence of contemptible credulity, erratic imagination or mental instability, the intellectual Establishment has not only ignored its own interests by neglecting a potentially rewarding field of study, but failed in its responsibility to the community.[38] The net result is that hundreds of thousands of men and women have had a strange experience, but have found that there is nobody to tell them what that experience was.

Or rather, nobody to tell them *authoritatively* and *scientifically* what that experience was. There are priests who will endorse one visionary's sighting of the Virgin Mary, while keeping quiet about the hundreds of almost identical cases for which the Church prefers not to vouch. There are spiritualists who confirm a sighting of the witness' dead relative, without mentioning the very good grounds there may be for believing that it was some kind of impersonation. There are ufologists who encourage witnesses to believe that they really did see extraterrestrial entities step out of a spacecraft from outer space, without drawing attention to the fact that we do not have a scrap of evidence for the extraterrestrial origin of UFOs. Many of these self-appointed amateur experts are knowledgeable, perceptive, well-informed, besides being sincere in their efforts both to understand the phenomenon and to aid the percipient. Nevertheless, their explanations are seldom better than subjective hypotheses, and frequently coloured by the explainer's personal belief-system. Yet, as things now are, the percipient in search of an answer is unlikely to do better elsewhere.

One of the reasons for the silence of official science is that nobody is quite sure which branch of science should take the responsibility. Many are willing to accept that a genuine phenomenon exists — but what kind of phenomenon is it? If you do as has been done so often in the past, and simply label the experiences 'figments of the imagination', you relegate the phenomenon to the domain of the psychologist, which at this stage of research could be premature, for it may well be nothing of the kind. We have to retain the option that the entities may be what they claim to be — spirits of the dead, extraterrestrial visitors, ministering angels or seducing demons. Many of the percipients, and many of those who have tried to interpret their experiences, have believed that some of the entities possess total physical reality. Unless and until we can prove that this is not the case, we do not have the right to discount this first-hand testimony.

For the purposes of this study, then, we must embark in the belief that a real phenomenon is taking place, though whether on the physical or the mental plane (insofar as those terms have any meaning) we shall not at this stage seek to decide. We shall, before we go very much further, have to agree on what we understand by 'reality'; for the moment, let us agree that whatever kind of reality the entities may or may not possess, the *reports* at any rate have reality. People like Glenda are making sincere statements about experiences which they genuinely believe themselves to have had, and that makes them undoubtedly real in at least one sense of the word; something occurred which impressed them, frightened them, delighted them, dismayed them, enlightened them, comforted them, alarmed them, reassured them or simply 'shook them up', according to its nature and their state of mind.

A preliminary overview

We shall be looking at these experiences from several different points of view. We shall be asking:

★ What is seen? What are its outward characteristics and what is the entity's apparent identity?

★ What does the entity do, and sometimes say?

★ When and where is the entity seen? Are any special conditions necessary for it to be seen?

★ Who sees entities? Are people who see entities a special kind of people, or are they, for that moment, in a special state? How do they react to their experience? How do non-percipients respond to their accounts?

★ *How* are the entities seen? What physiological and mental processes are involved?

These aspects of the phenomena will recur throughout our study, but a general overview of what we are likely to find, here at the outset, will help us to proceed in the most useful direction.

The *identity* of the entity — who or what it appears to be — is likely to be the most interesting aspect of it so far as the witness is concerned, and for the investigator it is the feature which offers the greatest opportunity for further study. It is also the feature which seems most clearly to distinguish between one kind of entity and another. We can therefore pigeonhole our sightings on this basis without too much difficulty:

1   At one end of the spectrum is the entity which appears to be a friend or relative, living and known to the percipient, but who appears in circumstances which, whether or not they seem perfectly natural at the time, are seen later to be physically impossible. For instance, it may be clearly established that the apparent was somewhere else at the time.

2   Next come sightings where the apparent is dead, or in the process of dying, at the time of the sighting. This fact may or may not be known to the percipient; but this does not necessarily affect either the form of the manifestation nor the way the percipient responds to it. (It is just this sort of paradox which can suggest further lines of inquiry: the fact that the percipient's conscious mind is unaware whether the apparent is alive or dead immediately tells us something about the role played by the percipient — a point we shall take up later.)

3   In the next category, the apparent is a stranger to the percipient. Later, the identity of the entity may become known to the percipient,

as in the 'haunted room' type of case where a guest comes down to breakfast and tells her hosts of a little old man who appeared in her bedroom, whereupon she is told it was the ghost of the seventh duke, or whatever. In other cases, the identity is not discovered, but of course this does not mean that it is not discoverable, simply that nobody is sufficiently well informed. Occasionally a diligent investigator may, as the result of his researches, be able eventually to confront the percipient with a photograph of a likely candidate, whereupon the percipient may exclaim, 'Yes, that's my ghost!' which is not worth much but is better than no evidence at all.

4   Next come those cases in which the apparent, though not known personally to the percipient, is easily recognized. This is notably the case with religious visions, where the percipient usually has no doubt that she has encountered Jesus Christ, the Virgin Mary, or whoever. This is not to say that the percipient cannot be mistaken. We shall see that such identification is not always reliable, however much the percipient may insist on it, and there are also instances in which the identity of the entity is not immediately obvious. A noteworthy example of this is the vision of Bernadette Soubirous at Lourdes, where only after several manifestations was the identity of the entity established. With this category, too, there is, even more than with other kinds of apparition, the possibility that the entity itself is lying about its identity; a favourite trick attributed to deceitful devils by medieval theologians was the assumption of the appearance of a saint, thereby to win the trust of the percipient.

5   Sometimes the apparent is not human at all, in an earthly sense. He may be an angel, a demon, or a visitor from some other world or plane of being. In such cases the percipient, with or without the help of his friends or advisors, has to decide who or what he is seeing, with of course every possibility of misinterpretation. The literature of religious visions is full of sincere witnesses being told that their heavenly vision is really a diabolical counterfeit. In such cases, everything depends on the prevailing climate of opinion and the personal prejudices of the visionary's confessor or counsellor. The case of extraterrestrial aliens is especially intriguing, for there are no experts to whom the percipient can turn with any confidence to tell him what he has seen: rare indeed is the psychologist or other practitioner who will diagnose anything but some kind of delusion. Consequently, it is left to the percipient himself. Extraterrestrial aliens are so frequently identified as such, despite the percipient's lack of experience in these matters and our present state of ignorance as to what an extraterrestrial alien actually looks like: what makes him so sure?

6   Finally, there are those cases where not only does the percipient not identify the apparent, but he does not *expect* to. In the case of hypnagogic images, for example, the percipient is likely to accept that

the nature of the entity is something quite outside normal experience, and he will not seek to identify the image with mortal persons, living or dead, nor with spirits or aliens. Instead, he simply resigns himself to the fact that the entity has no 'real' existence, and leaves it at that. In this he may or may not be justified; but certainly it would be wrong for us to accept this negative interpretation without question, just as it would be wrong to take the positive identifications at their face value.

Even in the course of making this brief analysis, I have had to indicate that the compartments are by no means as watertight as they seem: we cannot depend on any of these entities really being who they purport or appear to be. So, while it does indeed seem practical to treat the apparent identity of the entity as its single most fundamental characteristic, we must recognize the dangers of treating any classification based on identity as anything more than a convenient device. The dangers of doing so become apparent the moment we look more closely at any of those categories. Take the simplest, seemingly the most clear-cut of all, the apparitions of recognized relatives. We shall find that these take place in a wide variety of forms: sometimes the apparition will be so lifelike as to deceive the percipient into thinking that the apparent is actually there, at other times it will appear in a totally fantastic manner. Are we justified in lumping these two kinds of apparition together, when it would seem that the process of manifestation and perception may be quite different?

Similarly, there may be attributes of some cases in one category which are shared by some cases in another. Here is an example. There are many cases in which apparitions of friends or relatives appear and warn the percipient of some danger; but equally there are instances of similar warnings being given by quite different kinds of entity — by religious figures or even extraterrestrial beings. If we assume, as surely we must, that the giving of the message is purposeful, it either constitutes the reason for the entity manifesting at all, or it is at the least a deliberate action, as though the entity, here for whatever reason, chooses to warn the witness about some impending event. Either way, we cannot escape the implication that a whole variety of different kinds of non-human entity are, intermittently at least, concerned about the well-being of individual humans. This is not, of course, an impossible situation, but it is certainly an implausible one. This fact alone would justify us in looking for an alternative explanation, among which has to be the inference that there is not so much difference between an apparition of Aunt Agnes and one of the Virgin Mary, or Ashtar, as might at first appear.

If the identity of the entity is not a valid basis for classifying sightings, what about the process whereby the entity makes itself seen? Perhaps what seems to be a difference in kind is no more than a difference in means, just as an author may tell his story as a novel, a radio play, a film or whatever. There may be all kinds of factors which would cause the agent to choose one means of manifestation rather than another — inability to 'transmit' except in a particular mode on the agent's part, a similar incapacity on the recipient's part, or some external circumstances about which we can only speculate.

There are various ways in which the 'quality' of the entity may vary. The temptation, of course, is to assume that the manifesting agent is aiming at complete verisimilitude — that what he would like best would be to appear before the percipient so naturally, and so natural-looking, that the illusion would be perfect. This does seem to happen in some cases, where it is only subsequently that the witness realizes that he has seen an apparition. (From this it follows that many supposedly real people we encounter in the normal course of life may not be as real as we think: we may be encountering non-real entities many times a day without realizing it!)

But while it does seem, in the majority of cases where the illusion is less than perfect, that perfection was being aimed at but unfortunately not achieved — as when we only see part of the figure, or the legs seem to fade into nothing rather than extend into feet — there are also many cases in which the agent does not seem particularly concerned to achieve a natural effect at all. Sometimes, the entity manifests simply as a face, or a head and shoulders, in such a way — for example, framed in a piece of furniture — as to suggest that the agent had no intention of filling in the rest of the figure, and would not do so even if he could. It is the same with the movement of the entity. Some walk into the room like a human being, seeming to open the door, turn the knob, and so on, and once in the room move about it with strict regard for the furniture; others, on the contrary, suddenly manifest in the room without any indication as to how they got there, and are perfectly happy to move through the furniture with total disregard for being lifelike. It does not seem as though the latter type are failing to become the former: it is more as though the agent had other priorities, just as some artists do not care whether they achieve a photographic likeness of their sitter, so long as they express his character.

In the event, such cases seem no less 'successful' than the others. However unnaturally the event may have come about, the percipient can be persuaded that he really is in the presence of the apparent, accepting the ensuing experience despite the bizarre circumstances. Indeed, there appears to be the same kind of 'willing suspension of disbelief' as operates in a theatre: the percipient accepts the 'conventions', and does not insist on perfect verisimilitude. But even this we cannot be sure of. Although the percipient consciously assents to the manifestation, is disbelief wholly suspended? Perhaps not, if the behaviour of sitters at spiritualist materializations is anything to go by. Many commentators have noted the extraordinary equanimity with which witnesses will accept the seeming presence of a dead relative's return from the grave, an experience which ought to be psychologically shattering yet which most percipients seem able to take in their emotional stride, give or take a few sniffs of tearful joy. The inference is that perhaps there is a part of the percipient's mind which is acting as a kind of damper, permitting a conscious assent to the apparition's identity but withholding a total acceptance of the event in its momentous implications.

There are many other ways in which the mode of manifestation may vary, but the foregoing will serve as an example for our present purpose: we shall consider some others when we come to study the phenomena in greater detail. The point we should note at this stage of our inquiry is that there seems to be no obvious correlation between the identity of the entity and the way in which it manifests. Certainly it is true that spiritualist manifestations are all more or less unnatural in their mode of appearing — as is perhaps only to be expected since they are summoned up in a rather unnatural manner! But when they appear they do so in a great variety of ways, some forming gradually from balls of light and others from clouds of mist, some looking like cardboard cut-outs and others appearing as traditional draped figures in long robes, which may indeed be the prevailing fashion in the next world but inevitably strikes the sceptic as being admirably suited for impostors and quick-change artists. Visions vary from the more or less lifelike, as at Lourdes, to the picture-like, as at Pontmain, where the visionaries seem to have been treated to a kind of celestial *tableau vivant* which was added to while they watched.

Ghosts are in general more natural in appearance, and unlike spiritualist manifestations are willing to revert to their earthly style of dress, which often facilitates identification. But many ghosts are less careful about behaving naturally, and may indeed appear only from the waist upwards, for example. As we shall see, hallucinations are the most lifelike of all, manifesting with extraordinary vividness even though they seldom deceive the percipient into accepting them as real. But of none of these categories can we say that one kind appears in one fashion, another in a different fashion.

There is a similar absence of correlation when we turn our attention from the thing seen to the person who sees it. The circumstances under which a percipient undergoes an entity experience are about as varied as it is possible for them to be. Consider, for instance, the spontaneity or otherwise of the event. Probably the greater number of entity experiences are reported as occurring spontaneously: the percipient did not consciously cause or even wish them to happen. But there are many exceptions. An entity will often manifest in response to a wish or fear, whether articulated or not. For example, when the percipient is in some danger and can be thought to be sending out a non-directed call for help, an entity may manifest and give either comfort or assistance, which can be of a very practical kind. Sometimes the manifestation will be as the result of a deliberate attempt on the percipient's part: an exercise in astral projection, for example, or a magical rite intended to conjure up a spirit from the vasty deep. These exceptionally interesting cases provide us with our only means of making a direct comparison between those entities which appear spontaneously and those which are deliberately induced. The fact that both kinds of entity are of many different kinds, and that the kinds of some spontaneous entities strongly resemble the kinds of some contrived entities, is perhaps the strongest indication that any distinction between them is more apparent than real.

In the course of our study we shall find that there are many other characteristics which are shared by entities from different categories, but not necessarily by all entities in any one category. An example is multiple sighting. In several different categories there are entities which are reported as being seen by more than one person at the same time. Ghosts have been seen by several witnesses, simultaneously or sequentially; religious visions are frequently shared by a group of children; spirit manifestations are often seen by several sitters. It is also true that in all these categories there are cases where the apparition is *not* seen by some of those present. Whatever these two factors mean — and their implications are of course profound — it is likely that they mean the same thing, whichever kind of experience is concerned. If multiple sighting implies a degree of reality when the three visionaries of Garabandal see it, then it does so when a departed soul returns to the seance room. If, on the other hand, the explanation is mass or shared hallucination in the former case, then so it probably is in the other. And so on.

The normal process of perception, whereby we come to recognize that we have seen something, is a complex but sufficiently understood process. If our brain receives signals which indicate that there is something in a certain place, when we know that objectively speaking there is nothing in that place, then we know that an alternative stimulus is somehow being fed into the process in substitution for what should really be there. Who or what is doing the feeding?

The mind seemingly possesses a storehouse of imagery, or has access to some communal source, on which it can draw as required. Thus we can, at will, recall a scene once visited, a film once watched, a person once known. Side by side with the storehouse, the mind seems to have a workshop, in which this imagery can be arranged into patterns more or less meaningful. This can of course be done on the conscious level. It can also occur as a subconscious process, most notably in the form of dreams, but also as a consequence of other altered states of consciousness which have the effect of encouraging 'image games'. At times, as in experimentally-contrived states such as sensory deprivation, or in spontaneous conditions such as the pressure of strong suggestion or emotion, this material obtrudes into our conscious mind and may then masquerade as reality.

Most of the time we are sufficiently alert to subject any such material to a process of reality-testing, and we then decide whether what we think we see is real or illusory. But under certain conditions we are unable to do this — or, perhaps, some controlling power sees to it that we do not; it is then that we can confuse non-reality with reality. What is more, this is itself something that can be done either wittingly or unwittingly. As we shall see, it is quite possible for a percipient to know full well that the thing his brain tells him he sees is not 'really' there, and yet for him to go on 'seeing' it just the same.

How this alternative-stimulus occurs is secondary to the question of *why*, and in answering the primary question we may help to answer the other. It is in large part the object of this study to determine *why* people see entities. Why do some seem purposeful while others seem random or arbitrary? Is there an external trigger, even external control or manipulation, or is it purely an internal matter between our conscious and unconscious minds? These and other questions will recur, and our study of cases will suggest some answers. All we need to assert here at the outset is this: that *if something (it may be our unconscious mind, it may be some external force) wants to make our conscious mind believe it is seeing something even though that something is not objectively there, the necessary mechanism does exist and its physiological nature is known.*

The medical profession has long had to cope with patients who report visionary experiences which have no basis in physical reality, and so the phenomenon of hallucination has become a perfectly respectable one for doctors to study, though it is still not considered altogether respectable for people actually to experience such hallucinations themselves. But while there has been considerable discussion of the mechanics of the process, its causes and cures, this has all taken the form of treatment of an ailment. Wherever possible a physical cause has been looked for, and often it has been found. It has been demonstrated that the hallucinatory experience can often be attributed to such factors as dietary deficiencies or excesses, the symptomology of certain illnesses, or the side-effects of certain beneficial drugs. All of which is perfectly sound, and helpful as far as it goes. But it only goes as far as describing the kind of conditions which favour the kind of image-substitution process we have noted as providing a physiological mechanism for the entity-sighting experience: it does not explain why, when those conditions exist and when that process has been activated, it takes one form rather than another.

Yet it is this aspect which is in many ways the most interesting. Except to a doctor, it is not particularly interesting to record that when an individual is in a certain physiological state he will be more than usually liable to experience hallucinations; but it is very interesting to know that, say, if he is suffering from a certain ailment he will have one kind of sighting experience, whereas if he takes another type of drug he will have quite a different kind. We shall be considering some of the medical aspects of entity-seeing in greater detail later. For the moment we must recognize the importance of the witness's circumstances, physical and emotional, as contributory factors which may cause some, or all, of the experiences we are studying, or may simply put him into the requisite state in which others can cause him to have the experience.

It is evident that an entity sighting is a many-faceted event. It is characterized by who sees it, and under what circumstances; by the ostensible identity of the entity and by its apparent purpose; and by the manner in which the manifestation takes place. It follows that any attempt to classify such experiences according to any one of these parameters is liable to be at the expense of the others: and since we do not know which, if any, is the most important, we must do our pigeon-holing with caution.

Must we pigeon-hole at all? Clearly there *are* differences in kind between some sightings and others, and it would be as wrong to ignore them as to exaggerate their importance. The danger lies in assuming that a difference of one kind implies a difference of another, or that any distinction is as absolute as it appears.

There is one parameter we have not taken into account to any great degree at this stage; that is, who or what is *causing* the sighting? Clearly this is the most important question of all; for if there is one thing we all want to know, it is whether there are really non-human beings intruding, benevolently or otherwise, in our lives, and the study of entity sightings probably offers a better means of finding an answer to that question than any other line of research. This question will underlie our entire study, and we shall continually be relating our findings to it: for example, when we find an entity giving help to the witness, we shall have to ask ourselves whether we at last have sure proof that some benevolent being — an Earthperson who has moved on to a further plane of existence, say, or a discarnate spirit who never was a mortal being — is intervening on his behalf? But we must avoid making any premature inferences until we have considered the whole bulk of the evidence.

By way of reminding ourselves how important it is to keep an open mind when considering the evidence which follows, here are a few examples of scenarios which have been proposed:

★ The Devil is anxious to gain control of our souls, and to this end he masquerades as a benevolent spirit, or as someone known to us, winning our confidence and leading us to defect from our allegiance to God.

★ Aunt Agnes, now dead in the earthly sense and inhabiting a different plane of existence, is bothered by something unresolved from her earthly life, and so returns to Earth, either haunting the place where she once lived, or manifesting to a relative with a message, or taking advantage of a spiritualist communication system.

★ A child, starved of comfort, imagines a companion who befriends and advises her. So strong are her wishes that the fantasy-friend takes on a degree of apparent reality and even acts autonomously.

★ Extraterrestrial creatures are monitoring events on Earth, and choose selected individuals whom they contact and appoint as their emissaries.

★ The Virgin Mary, mother of Jesus Christ, concerned for the welfare of mankind, returns to Earth. The witnesses she chooses are humble, simple people, usually children uncontaminated by sophisticated ideas. She utters enigmatic messages to test their faith, and gives warnings on both a personal and a global level.

★ A person in a dangerous situation is rescued by a mysterious being who suddenly materializes, seemingly from nowhere, and saves him. Then, when the danger is past, he vanishes equally mysteriously.

★ A group of people, hoping to make contact with the spirits of their loved ones now passed on to a higher life, find themselves visited by the form of someone whom they recognize visually and who speaks and reacts in a way they identify as characteristic of that person when living.

★ A person on her deathbed finds herself surrounded by figures whom she recognizes as friends and relatives who are already dead; they may include people she believes to be still living, but who are later found to have been dead at the time. These figures seem to be 'welcoming' her into a new sphere of existence.

★ A man is thinking so hard about a friend, that a part of him — a kind of duplicate body visually identical to his physical body — detaches itself and travels to where the friend is, and is seen by that friend. Moreover, this double acts as though it is aware of its surroundings, even though its physical counterpart has never been there.

You may consider that some or even all of these are ludicrous interpretations of what is taking place. However, apart from the fact that each of them is backed by a strong body of belief, each is a not unreasonable way of accounting for the reported facts. If we reject them, it must be either because we can demonstrate their intrinsic impossibility — prove, for instance, that there is no Devil, or no Astral Self — which is itself a wellnigh impossible thing to do; or because we can offer a more plausible alternative, better supported by the evidence.

To that evidence we are at last almost ready to turn, but first there is one knobbly question which, if we do not tackle it now at the outset, will bother us throughout our study, it is: What do we mean by reality?

Most of us, most of the time, are aware of perceiving things in one of two modes. On the one hand there is what we term 'reality' — the touchable, smellable, matter-of-fact world of everyday life, which we know can include some very improbable things, from ball lightning to moon-dust, but whose 'reality' we do not question. On the other hand is what we are apt to term 'dream' or 'fantasy' or 'imagination' — a kind of non-existent parallel universe, which we tend to think of as having been invented inside our own heads.

Most of us, most of the time, believe ourselves capable of distinguishing clearly enough between these two modes. We know that there are times, such as when we have diminished the efficiency of our brains with drink or drugs, or when this has been done to us by illness, when we are not capable of making reliable distinctions: but we accept this as exceptional.

Indeed, we are so confident of being able to draw clear-cut distinctions that we can amuse ourselves by pretending to blur them. For instance, we can pretend that reality is not real, or that unreality is. Watching a television programme, we know that the projection itself is real enough, consisting of the manipulation of known physical substances by known electrical forces; we know that a real cameraman filmed real actors in order to create the programme, though what we see on our screen is only a record of what they once did; we know that, though the actors were moving in a three-dimensional environment, we are seeing it in only two dimensions; and further, we know that though we go along with their claim that the place they are moving in is the castle of Elsinore, it is only a simulacrum of a castle hammered together in a television studio, and not a real castle at all, and even if it was a real castle — if the programme was made on location — it would still not be Elsinore, because Elsinore has no actual existence, was never more than an imaginative creation of someone else's mind.

Clearly there is ample scope for confusion here, and we ourselves will on occasion have looked at the screen, especially on the 1st of April, wondering if it was fact or fiction we were watching. But such confusion does not generally worry us, indeed we tend to enjoy it; and it certainly does not diminish our conviction that the distinction between reality and non-reality exists and is an absolute one. Moreover, we regard that distinction as important, even essential, to our response to life, the universe and everything. However enjoyable, astonishing and sometimes revealing the non-real world may be, it is nevertheless 'no substitute for the real thing'. We may not always like reality; we may go to a lot of trouble to escape from it into a world of illusion; but there is never any doubt in our minds (unless we lose control of them) that reality is more 'important' than non-reality. (So much so that if we happen to prefer the non-real world of our imagination to the real world, we will do our utmost to make it real, with a variety of results, including a myriad transcendental communities turning their backs on modern civilization. This urge towards realization of our unreal dreams is a factor which may well turn out to be relevant to our inquiry.)

In most contexts, our criteria for determining whether a thing belongs in the real world or not are clear and trustworthy. Most of us, asked to supply an example of what we understand by reality, would start with our own selves. 'I think, therefore I am' was where Descartes started, and, consciously or otherwise, that is the starting-point that most of us would choose. From there we would move gradually outwards, via the chair we are sitting on to the house we live in, and so on. 'I can feel it, therefore it is' may not be scientifically or philosophically viable, but it works pretty well in practice. Most of us now know that what we are touching is 'really' a nearly empty space, populated only by a few wildly careering micro-units of matter, and that even those units might be better described as a force; but we do not therefore dismiss ourselves, our chair, this book, as illusion. We know that what our senses report is a convenient version of reality, that is all. For practical purposes we find the reports of our senses to be a reliable enough guide so long as we make due allowance for misinterpretation: and for more specialized purposes we have devised instruments capable of extending the range of our senses, to give us a more scientific account should we require it. A memorable advertisement caption from a few years ago reminded us that 'THE GIRL YOU LOVE IS 70 PER CENT WATER'. We accept that this is true, even though we cannot confirm the statement with our senses; but since most of the time her chemical constituency is not what is most important to us about her, the more superficial account presented by our senses is generally adequate.

On the other hand, our fantasies about rescuing her from danger and earning her undying gratitude, of winning her love by doing some splendid deed — these we regard as non-real in any sense. Dreams represent the epitome of non-reality, and into this category we would place a whole range of mental activities. Asked to supply an example of what we understand by non-reality, it is a dream or a fantasy that we would choose. Yet, as we shall shortly see, the non-reality of dreams is not quite as absolute as we generally assume: to take the example above, if our conduct is affected by our wish to win the loved one's affection, if it modifies our behaviour and perhaps affects the entire course of our careers, can we meaningfully say that it was a non-real cause that brought about these very real effects? Just as the physical world involves a great deal of illusion, so the world of illusion has physical aspects which raise disconcerting philosophical questions.

If, when speaking of solid tables on the one hand and insubstantial dreams on the other, we have to qualify our use of the terms 'real' and 'unreal' by 'it depends what you mean by . . .', this is even more the case with those phenomena or happenings which are neither as solid as a table nor as elusive as a dream. Phenomena which are not clearly related either to the real or the non-real modes of being have perplexed humanity ever since it started leaving any record of its thoughts: and because nobody has ever worked out quite what to do with them, they have been relegated to a kind of pending file labelled 'the paranormal'. (It has had different labels at other times and in other places.) Periodic attempts have been made to clear this pending file by assigning its contents to either the real or the non-real world, but nobody has ever managed to do so satisfactorily except in respect of a few minor items. After more than two thousand years, ghosts and poltergeists and visionary beings of all kinds remain in the pending limbo, a perpetual reproach to us. Frequently our intellectual leaders have sought to insist that 'there is no such thing as a ghost', but those who continue to see ghosts have not been persuaded that this is so. On the other hand, when the spiritualists offer a matter-of-fact scenario that establishes ghosts firmly as spirits of the surviving dead, as real-life as you or me though operating now on a different plane, this in turn is rejected by the majority. And so ghosts remain in the pending tray, labelled but ill-defined and certainly not explained.

The subject matter of our present study, of course, falls entirely within this debatable ground, and the question 'Is it real?' was probably asked, of himself or of others, by the percipient in every one of the cases we shall be examining. Implicit in the fact that this study has been undertaken is a broad 'Yes' in answer to that question: an affirmation that these experiences, reported so widely over so long a period, do possess reality of a kind. And further, that it is a kind of reality that is susceptible to study, that can meaningfully be analyzed, classified, compared. And that for these reported phenomena, understanding of their processes and identification of their causes are ultimately discoverable, even though we ourselves may not be fortunate enough to discover them.

The best way to be sure that a thing is real is to make it yourself. When a conjuror performs a trick before you, you know that your eyes are liable to be deceived: when an entity manifests before us, few of us will dare to put total trust in what our senses tell us. But it is another matter when we ourselves have performed the trick or caused the entity to manifest. So in the second part of our inquiry we shall feel ourselves on relatively surer ground, as we consider a range of experiences in which control over the phenomenon is, to some degree at least, in the percipient's hands.

However, the notion that there is a distinction between spontaneously-appearing and deliberately-contrived entities is one that can be dangerously deceptive. For the ones that we did not conjure up ourselves must have originated *somewhere;* someone or something must have brought them into being. But at least, when we pass from those entities that appear to manifest spontaneously to those we know we caused to manifest, we have eliminated a part of the uncertainty; and if we then find that there are similarities between the two kinds, it will be a fair presumption that the process whereby the contrived entity came into being is similar to that whereby the spontaneous entity came into being, and that whatever degree of reality the one possesses, the other is likely to possess in equal measure.

The purpose of this tiresome note has been to make sure that none of us proceeds further into this study without discarding assumptions which might impede our understanding. We are about to enter a realm where reality is not quite the same as the kind we are used to; but let none of us suppose that in doing so we are entering a Wonderland where reality no longer exists or has meaning. *These experiences we are about to examine were real experiences, and whatever caused them was real also.* If we do not end up with the correct answers, it will not be because those answers are essentially unknowable, but because we have not been clever enough to ask the right questions.

PART ONE

THE ENTITY EXPERIENCE

1.1 ENTITIES IN DREAMS

The dream, in our everyday speech, is a synonym for unreality. When we say of a thing that it was 'like a dream' we mean that it was as different as can be from our habitual waking experience. Even though, while in the act of dreaming, we become very involved in the action — frightened, disturbed, delighted — we are aware, either at the time or when recollecting it, that this difference exists; and we tend to feel that the difference is an absolute one, dream world and real world being separated by a border as clear cut as the looking glass through which Alice passed to her Wonderland.

The main reason for this feeling is, of course, the difference between the way things happen in the dream world and the way we are accustomed to them happening in the real world. Dreams continually defy the logic of everyday life: we move instantaneously from one place to another, we encounter people we know to be dead in real life, we are able to do superhuman things, the normal laws of cause and effect seem to be suspended. But at the same time our response to these peculiarities is quite different from what it would be if they occurred in real life. We accept the bizarre situations with equanimity, and can often exercise a fantastic degree of control over the dream-experience — getting out of dangerous or embarrassing situations by simply willing to do so, or bringing the dream to an end like an adult halting a children's game when it gets too rough. Moreover, while a dream may possess a very strong emotional cast — often more vivid than a real-life experience — we are able to feel a certain detachment from this even while being affected by it; depressed by a dream, we are consoled by our knowledge that it was 'only a dream'.

Of course the pre-eminent unreality of the dream is that it is such a solitary experience: except very rarely, and then only partially, we cannot share our dreams. The most we can do is to tell other people about them next morning at the breakfast table. Yet it is this characteristic which, paradoxically, is the surest evidence we have that the dream does possess some kind of reality; for the very fact that nobody else could have experienced it, coupled with our ability to tell people about it, means that it must in some sense have 'happened'.

And as soon as you accept that, you acknowledge that dreaming is a biological reality, something that we were born with the ability to do, a capability that was built into us by whatever forces are responsible for our existence and design, something that the biologist must be able to structure into his account of the human being and whose function he should be able to discern. If he should happen to be entrenched within the behaviourist position, it is all the more incumbent on him to determine what purpose dreaming serves.

That dreaming does have a purpose is of course a widespread notion, held throughout history. In the most ancient cultures, as in many primitive societies today, this was so generally accepted as to form part of the prevailing religious doctrine: dream interpretation features notably in such records as the Old Testament, for example, and also in such societies as the Senoi of Malaya. [154] Such metaphysical notions have not of course been generally acceptable to science, which has tended to relegate them to the pending tray of the paranormal, along with magic and witchcraft, as a marginal aberration of human behaviour for which there is presumably some natural explanation. Pressed to suggest what sort of natural explanation they have in mind, scientists have proposed that perhaps dreams serve some kind of biological purpose by acting as a refuse bin into which the brain can discard the unwanted rubbish it has accumulated during the day, like a computer effecting clearance of redundant programs

Meanwhile cooks and housemaids, and many others, have continued to look for meaning in their dreams, like Joseph and the other noted dream-interpreters of history: and towards the close of the nineteenth century Freud and others justified their faith by showing that genuine and useful information could indeed be gleaned from studying the content of our dreams. It did not necessarily follow that dreaming was henceforward to be seen as a positively creative activity: just as clues to a murder may be found by the police when they rummage through the waste-bins in the suspect's apartment, so the retrieval of useful information from dreams does not necessarily raise their status. It is still an undecided question whether dreaming should be regarded as a purposeful process or as a kind of mental excretion.

However, even adopting the most minimal explanation presents us with challenging questions, for it is evident that the process of dreaming involves a great deal of very elaborate mental activity. While it is possible that this is to be compared to a very sophisticated waste disposal system, it is not easy to accept that our unconscious minds would put themselves to so much trouble simply to relieve our mental systems of so much unwanted matter. We do not, after all, compose our litter into works of art before discarding it; yet that, on the rubbish disposal hypothesis, is what our subconscious minds do. Rather than endorse such a 'spring-cleaning can be fun' explanation, it is easier and more satisfactory to hypothesize that dreams are the result of some biologically useful mental process which has some higher, possibly even a vital, role to play.

That much of our dreaming does have a positive function seems certain, then, and justifies us in treating the dream experience as a 'real' activity and the dream material as, in its own way, real. Which encourages us to recognize the various ways in which the dream-experience is not quite so far removed from the real world as habit — or science — has persuaded us to think. To start with, there is the fact that the component parts of the dream are taken from our daily lives and that past and present are inextricably mingled as though time was irrelevant. But then we find that the same applies to the future: precognition of coming events occurs so frequently in dreams that only our reluctance to admit 'what the dream said' as evidence prevents us from accepting that the 'dream-producer' appears capable of moving as freely as he chooses in time as well as space, with all the implications that this entails.

Such a line of inquiry is outside our present study, except that it is an important indication that *whatever creates our dreams has access to a far greater range of material than our conscious minds appear to enjoy:* and this, as we shall see, is going to be of the very greatest importance for our study.

Who/what our dream-producer is, must be a supremely important consideration. That our dreams are created in the unconscious mind is generally assumed; and it is further assumed that this implies that they are created *by* the unconscious mind. It may well be so, though we should do well to remember that the unconscious mind is not a scientifically observed objective fact, but simply a label for something we feel probably exists. For the purposes of our study, we may as well go along with this familiar and convenient assumption; but let us not neglect any clues which may tell us more about it.

One such clue is apparent to us all: the fact that our dreams are to some extent regulated by external factors — our state of mind, our health, our anxiety level, as well as by noises and other sensory stimuli, suggestions from other people whether spoken or thought, and many more. Which is to say that external events, occurring in the real world and exerting real effects, are modifying our dreams. Here, right at the outset of our inquiry, is a demonstration of an interface between the real and the supposedly non-real; and of the indeterminate nature of the boundary between them. For it shows that the dream-producer is continually monitoring reality even while busy with the job of creating our dreams, and that he is ready to make use of whatever material comes to hand, often with marvellous ingenuity as well as breath-taking speed, providing an elaborate scenario, highly detailed, to plausibly 'account for' the sound of an alarm clock or the sensation of cold resulting from the clothes slipping off the bed.

Such game playing may not be significant, though it leaves the behaviourist with the task of explaining why we have, built into our systems, an instrument that is prepared to go to so much trouble for something that is not significant. But the dream-producer's work can take even more curious forms. For example, there are a number of people — I am one of them — whose dreams never, on any account, image the real world. Though I dream I am back at my Cambridge college, the Cambridge I dream about is *never* the real Cambridge; and so with any other place I dream about. Instead, my dream-producer goes to an immense amount of trouble to create a fictitious Cambridge, highly detailed, with fictitious streets and fictitious secondhand bookshops stocked with fictitious books, and fictitious rooms for me and my fictitious friends to live in . . . and my dreaming self finds its way round this counterfeit just as my waking self knows its way about the real Cambridge. Why, one has to ask, doesn't the producer make use of the existing real-life Cambridge, which after all lies ready to hand in my conscious mind? It all suggests a deliberate decision on his part, and yet his purpose utterly eludes me.

My specific puzzle is of course no concern of this inquiry: I cite it simply as an illustration of the way in which the unconscious mind is capable of manipulating reality, or substituting its own version of reality, as and when it chooses. Moreover, it encourages us to think that this is done by design: for I must emphasize that my unconscious is absolutely consistent — it *never* uses ready-made material, it *always* creates its own. And this ability does not seem to have been imposed upon my producer, for when I compare my dreams with those of others, I find that they generally have quite different adventures involving 'real' people in 'real' places: so a choice seems to have been exercised.

This ability of our unconscious minds, and the way in which that ability is used, means that here, at the very outset of our study, we have established the existence within us of *an autonomous creative instrument, the limits of whose capability we cannot even guess at, and whose motivations are no less unknown.*

It is at this stage that we must raise the possibility that we are wrong to think that our dreams originate exclusively within our unconscious minds. An alternative hypothesis exists: that they are scripted in whole or in part by some external source, which for reasons of its own plays them over to us while we are in an appropriate mental state (of which dreaming may be only one of several). The nature of the external source is, for the most part, something about which we can only speculate; but tutelary spirits, benevolent guardians and the like are the kind of thing we might hypothesize. (On the whole, it is not unreasonable to suppose that such influences would be benevolent rather than otherwise, for we may reasonably premise that such interventions are 'by invitation only', and that our dream-producer would, like the maitre d'hotel of an exclusive restaurant, refuse admission to an undesirable guest.)

I know of no way in which such hypotheses can be refuted, but clearly they present us with an immensely cumbersome alternative. They imply a totally unknown force, for which the evidence is circumstantial at best, that is capable of doing all that our subconscious minds are supposed to do, yet from outside not inside. It would not be worth giving consideration to such a notion, except as science fiction, were it not for certain curious reports which crop up from time to time, which assert that precisely this kind of intervention is taking place.

One intriguing class of such reports derives from reincarnation research. In the course of his work in this field, Ian Stevenson has encountered the occurrence of 'announcing dreams', which prepared the 'host family' to accept the idea that a child subsequently born to them was in fact the reincarnation of the deceased person appearing in the dream. Some time later, when the child has arrived and reached an age when it is able to claim reincarnation, it will claim also to remember — during the time after its previous death and before its present birth — sending an announcing dream to its new parents, or in some cases manifesting as an apparition. [153]

Only a few cases of this kind are known, and the evidence for some is little better than hearsay; but a single good case would establish the possibility of the phenomenon — and Stevenson is able to offer at least one case for which the evidence is very good. Now that he has been alerted to the occurrence of such phenomena, he and other researchers in the field will unquestionably be looking for confirmatory cases to provide a valuable secondary category of evidence for the reincarnation hypothesis.

Anecdotal cases in which dreamers are visited by deceased relatives and friends are of course nothing new, but for the most part commentators have not felt the need to suppose that the apparents were in fact responsible for their manifestation. If, however, reincarnation research should establish the reality of such intervention, perhaps we should re-examine some other cases, even though alternative explanations exist.

Another very different kind of instance in which external intervention in dreams appears to be indicated is provided by the case with which I opened this inquiry. The witness, Glenda, had a 'bedroom invader' type of experience; subsequently, the entity manifested in her dreams. As so often, we can choose between several possibilities. The dream-experiences could have been her subconscious mind making use of the previous experience, the 'visit'. Alternatively, it could be that it had been her subconscious mind that had created the entity in the first place and that was now serving it up in a different mode — perhaps for no other reason than that it happened to be more convenient.

However, if we can bring ourselves to entertain the notion that Glenda's visitor may have been a real extraterrestrial entity, an intriguing model suggests itself. We may suppose that Glenda was taken by surprise in the first instance, but subsequently put up some kind of mental block so that the entity could not appear to her again in her waking state; it then took advantage of the opportunities presented by the dream state, manifesting to Glenda when her defences were down. This presupposes that in the dream state we allow, as it were, the vacant premises of our minds to be used by whoever/whatever chooses, though, as suggested earlier, some kind of control may be exercised.

The hypothesis that some of our dream-experiences have an external origin is one which has been entertained by speculators ranging from primitive tribesmen to sophisticated occultists. It fits in neatly with many esoteric doctrines, notably that of the Theosophists, which we shall be considering when we look at folklore entities; it also provides the basis of one of the most constructive of all proposals for solving the entity enigma, the 'induced dream' hypothesis, which will get a chapter to itself in our concluding analysis. At this stage, however, the notion of an external origin for dreams must be regarded as little more than speculation until such time as they can be shown to be beyond the capability of our dream-producer.

The dream process is important for our study not simply because it produces dreams, in which entities are frequently seen, but also for the light it throws on mental processes in general. The dream state is not a totally distinct phenomenon, but the end of a spectrum that includes a variety of alternate states of consciousness (we have noted that dreaming is by no means the wholly unconscious process it appears to be, and this is supported by the physiological fact that dreaming occurs during phases of 'paradoxical sleep' in which the brain is as active as during the waking state, displaying a beta-wave pattern as compared with the gamma waves associated with the diminished activity of non-dreaming sleep states). While it would be premature to suggest that all the entity-seeing experiences we shall be considering are similar to the dream-experience, to the degree that what is true of one must be true of the others, it would be equally premature to think that we can leave the dream-experience out of our calculations.

A significant example of the relevance of the dream-experience emerged from Morton Prince's classic study of a divided personality.[121] We shall be looking at this very illuminating case later in our study; but it is appropriate at this stage to note that one of his patient's secondary personalities was able to tell him that her sleeping self would 'imagine all sorts of things . . . if she remembers them you call them dreams, and the others you don't'. The implication is that while one part of the mind is asleep and dreaming, another part is on the watch; and this leads to a further possibility.

We have a tendency to look on our dreams as 'events' in our lives: that is, we feel that from time to time we 'have' dreams, the implication being that each dream has been specifically prepared for that occasion. While this is no doubt true to some extent — for many of our dreams are clearly related to the immediate circumstances of our conscious life — this concept may nevertheless be giving us a false picture of the process involved, and I would therefore like to propose at least the possibility of an alternative way of considering what may be going on.

It is perhaps best done in the form of an analogy. A man comes home from work, switches on his television set and watches the programme that happens to be on. He will think of his television experience as what he happened to see because it happened to be on; the consequence of it is that the programme becomes an event in his own life, giving him memories, information, even perhaps an emotional experience which could conceivably affect the course of his life. Yet if he had stayed at home all day, he could have watched television during the afternoon and had quite a different experience; so too would it have been if his wife had video-recorded an afternoon programme for him to watch in the evening. She might even have played a trick on him, set up the system so that when he switched on what he believed to be the current real-time transmission, he was actually seeing something that his wife had prepared for him. Either way, a different event would occur in his life, giving him different memories, different information, a different emotional experience.

The analogy is not wholly exact, because whether he watches live television or video is likely to be a conscious choice, whereas our dream experiences seem to be wished on us whether we like it or not. But it holds good to this extent, that we might do well to think of the dream state not as a specific experience, but as a 'locking-in' to what is in reality a continuous performance, and to look upon what we experience and remember in that state as almost accidental so far as our conscious minds are concerned, however contrived and deliberate they may be on the part of those who produced and presented the dream.

This is not an idea plucked out of the blue, but is supported by the experience most of us have from time to time in which, in a state of reverie, we become aware of images flitting through our minds — images which, if we concentrate our attention on them, which we generally do not, we find to be astonishingly vivid and detailed. To speak again of my own experience — it is, after all, the one I know best — I know that I have from time to time found myself, for what is only a moment, experiencing a briefly glimpsed scene that is emotionally involving as well as visually detailed, yet at the same time clearly utterly imaginary: the effect is more than anything like seeing the briefest of film clips, a fully-realized scene yet one which seems to have no previous or subsequent existence, no discernible context, no frame of reference or meaning. As for purpose . . .

It is possible that these images are precisely the same as those which make up our dreams: alternatively, it may be that they are the undifferentiated material our dream-producer will manipulate and fashion when he comes to create a dream for us. But the important thing is this: perhaps our subconscious minds are *continually* projecting material, of which our conscious minds are only *intermittently* aware.

Several others have speculated along these lines. Professor H. H. Price wrote: 'It is plausible to suggest that image-thinking is going on in all of us all the time . . . that we are dreaming all day long as well as at night, but only notice it when we are asleep'. [120] At present we cannot regard this as anything more than a conjectural possibility, but it may help us in the study that follows if we think of our dreams not as one-off episodes but as brief glimpses into a continuous process. Again, this requires us to think of dreams as just one in a spectrum of modes of perceiving: as indications of a mental process that takes place within the normal personality, whether continuously or intermittently, to which the creative unconscious turns, we may suppose, whenever it seems to offer the most convenient vehicle for the story it has to tell, the information or warning it has to convey, the emotional impact it wants to impose.

From this it is only a short step to the conjecture that an entity seen in a dream may be no different from one seen in another category of experience. Such speculation, as we have seen, rests on a number of unsubstantiated suppositions; enough for the moment if I have shown that the possibility is there, for that would justify this brief journey into the world of dreams, which might otherwise have seemed an unnecessary detour. I hope I have shown that it is not so; that dream-experiences cannot be left out of consideration and may prove of the greatest significance for our study. As E. R. Hilgard has reminded us[70]: in dreaming, one part of us is spectator to the drama which another part of us is staging. Throughout our study we shall need to consider whether the same may be true of *all* the phenomena under our consideration.

1.2 ENTITIES SEEN BETWEEN SLEEPING AND WAKING

When falling asleep, or when waking, many people have visual experiences of a most extraordinary kind. Though not apparently universal, such experiences are certainly widespread: it has been claimed that one person in five experiences them, but the difficulty of establishing the fact make this a very unreliable figure. What is more important is that the experiences have in common features that present many of the problems also presented by other entity-seeing experiences.

Theoretically, there is a distinction between the imagery seen while falling asleep (hypnagogic) and while waking (hypnopompic): some people experience both, some only the one, some the other. But the indications are that this is merely a personal matter, and that there is no fundamental difference to justify our treating them as two separate kinds of phenomenon. (For this reason I shall from now on use the word *hypnagogic* to include hypnopompic imagery also, unless specifically indicated.) On the other hand, the kind of experience people report makes it quite clear that we *are* justified in regarding this kind of imagery as a phenomenon in many ways distinct from other kinds — from, say, sleeping dreams on the one hand or waking visions on the other.

The hypnagogic experience is probably the most elusive kind of image-seeing there is. Not only is it never shared by others, but the images flash across the mental vision of the percipient with such rapidity, and at a time when his intellectual forces are particularly inactive, that the experience is so fugitive as to be virtually unseizable. Despite the fact that we have all been doing it for so long, we still know surprisingly little about sleep: but it is evident at least that it represents some kind of altered state of consciousness in which our minds are working in different ways than when they are in the waking state. How different sleep is from other altered states of consciousness is a question that is currently being explored, but again the indications are that it has its own unique features. It is not, for instance, the same as hypnotic 'sleep', nor the kind of trance into which the practitioners of yoga, or psychic mediums, or religious mystics are wont to pass.

Whatever the nature of sleep, it is different from waking, and the process of passing from waking to sleeping, or sleeping to waking, must involve a kind of shifting of gears, as it were, or a changing of the guard, or even a decompression chamber that enables us to pass without trauma from one state to the other. Whichever of these analogies is the most appropriate, the effect during the process of transition is to cause a great many of us to have very special experiences. Once again we come up against the same puzzling questions we had to ask with regard to dreams: what biological purpose is served by these experiences? Why has the body been so constructed that during this transition state it will have these experiences? Why should these experiences be of this very specific kind?

Hypnagogic imagery has neither the 'real-life' clarity of consciously perceived images, nor the somewhat distant 'story-book' quality of a remembered dream, but rather a rapidly moving ephemerality in which our minds are not in a fit condition to record the experience with much precision. However, sufficient material has been collected by researchers to enable some statistical findings to be tabulated, and in 1925 F. E. Leaning was able to establish some fundamental characteristics of hypnagogic imagery[91]:

1   The material manifesting is usually unfamiliar. His witnesses stated that about two-thirds of them failed to recognize what they saw in this state:

	per cent
All matter quite unrecognized	60.5
Largely unrecognized	5
Usually or often recognized	20
Partially recognized	14

2   Virtually every one of his respondents remarked on the lifelike nature of the images. A high proportion used some such phrase as 'more vivid than real life'. The nature of the subjects seen was varied, but the most frequent were faces and landscapes: occasionally complete scenes were represented. When faces were seen, they were like those of real people, and often registered very extreme emotions.

3   For the most part, Leaning's percipients had no control over what they saw — they could neither cause a particular image to appear, nor influence its behaviour once it had appeared. Fifty-five per cent said they had no control whatever: most of the others said they could only occasionally exert some control. Only 14 per cent claimed they could exert 'much' control.

By and large, then, what most of us have by way of a hypnagogic image-seeing experience consists of the visualization, quite unexpected and involuntary, of a face or a scene, generally unknown to us and having no obvious significance, not therefore charged with strong emotion, generally — though by no means always — pleasant in character.

While accepting this as the 'norm', however, we should be aware of the possible variations occasionally reported, for we must not allow ourselves to be tempted into thinking that this kind of imagery is necessarily of a certain, limited kind. Here are some variants quoted by Leaning:

★ One percipient said the image reacted to him. 'I was resting on the sofa after dinner, with my eyes shut in the full glare of the electric light, when a picture was presented to me of a young cow, chestnut coloured, with white points . . . Behind the cow a small calf was standing. The cow seemed to sense my presence, and unmistakably adopted a protective attitude. Though I have been associated with cattle all my life, I am quite positive I had never seen that cow before.' and again: 'Another vision was that of a small species of duck unknown to me. It seemed to turn its head up towards me with an intelligent expression of inquiry and interest in its little black eye.'

★ Another percipient saw a detailed scene. 'Facing me there suddenly appeared a brilliantly lighted panel through which (or in which) I saw a room with three English officers bending over a map spread out on a table, and evidently planning an attack or movement of troops. The officers were moving as they conversed and pointed out places on the map: they had their backs to me, and I was panic-stricken lest they should turn and see me.'

★ Another reported hearing voices as well. 'There was quite a company of people about me, young women I believe, who looked toward me and passed on. One of them spoke. I heard the voice distinctly, soft and clear. It said "He isn't asleep." That is all . . . I am certain I am not confusing this with a dream.'

Because the outstanding characteristic of hypnagogic imagery is its unfamiliarity, combined with its involuntary occurrence, most of those who have sought to account for their experiences have tended to look towards some external source. One of Leaning's percipients thought she was seeing glimpses of past lives; others have wondered if they were seeing scenes from their own future lives; yet others wondered if they were witnessing actual people and scenes, but in some distant country. Leaning quotes the American researcher H. B. Alexander, who reasoned that 'it is self-evident that not being memories, nor things seen or previously imagined, they are mental constructs . . . the work of a highly differentiated mental compartment, without any apparent connection, emotional or volitional, with the aims, interests or feelings of the person concerned'.

If this is indeed the case, then it only highlights the puzzle of what purpose is served by the phenomenon. If the hypnagogic imagery originates inside ourselves, what is its biological function? If, on the contrary, it originates externally, why is the significance and relevance of the material so obscure?

The baffling nature of hypnagogic imagery is seen at its most extreme (and therefore, potentially, its most revealing) form in a particular kind of experience, which is quite frequently reported of this type of image-seeing but not, so far as I am aware, of any other. Typically, it takes the form described by the writer of this letter:

For many years I have experienced 'faces' and 'visions' of a sort, at that moment when I hit the borderline between waking and sleeping. It does not happen every night, but it happens often enough to make me wonder about it. At times I see a single face: sometimes masses of them. Sometimes I get a feeling of pure terror from them; sometimes they seem beneficent in nature. Sometimes they are 'human'; sometimes they seem nearer to non-human or sub-human. (Some, in fact, closely resemble faces of fauns and satyrs in paintings by the great masters. Is that where they got their 'models' for those faces, perchance?)

These things are not the product of my imagination. They seem to 'appear' only at that moment of 'crossing over' into sleep, and the 'vision' (I don't know what else to call it) seems to be projected onto my closed eyelids. Often they seem surrounded by such a sense of menace that I jerk upright into complete wakefulness. Although this does not occur nightly or even weekly, it occurs often enough to cause me to dread going to sleep.

Recently I experienced a slightly different version of the phenomena. In this case the 'vision' was not against my closed eyelids but beside my bed. I both 'sensed' and 'saw' it. It was a big dog sitting beside my bed, with his snout raised to the heavens as though he were howling. It was so close to me that I could 'see' every individual hair in its soft underthroat and it was almost near enough to brush my cheek.

This did not frighten me but has made me extremely curious; first, what is going on anyway, and second, what did that howling dog betoken, if anything?

I. M. Nelson, Milwaukee, Wisconsin[42e]

I can personally vouch that this sighting of faces is the most characteristic form of the hypnagogic experience. For me it takes the form of a procession of faces passing, fairly swiftly, before my eyes. For others the process takes somewhat different forms — often, for example, they are described as slowly materializing and slowly fading again; in my case the experience is more than anything else like a police witness watching a succession of criminal portraits flashed on a screen before him. But I certainly confirm the astonishing lifelikeness of the faces: they are presented with such clarity, such detail, such distinctive character, as to leave the percipient with the overwhelming impression of looking at real people, or at photographs of real people. So much so that I have wondered whether what I am seeing is my subconscious mind's file of all the people I have ever encountered in my life, sat next to on buses, seen across the room at parties, asked the way of in strange towns . . .

At the same time, the almost universal agreement — and I again confirm this — is that the faces are unfamiliar. Leaning quotes two pages

of statements to this effect: 'Always faces of people unknown to me
. . . always new, never those of friends . . . on no single occasion a
known face . . . strikingly distinctive, but never remembered'.
Is there any purpose behind their appearance? Canadian author
Victoria Branden, in her thought-provoking book on ghosts, [11] describes
a particularly alarming hypnagogic image:

> As I was drifting into sleep, the usual succession of images was flowing through
> my mind. Suddenly out of the misty landscape a motorcyclist rode, wheeled,
> and turned towards me. He was two-dimensional, part of the film, so to
> speak; the difference was that he could see me. He shook his fist at me, and
> I could see that he was mouthing threats. He seemed to be ordering me to
> get out, to go away.

Interestingly, Victoria Branden later had another sighting of the same
figure, but this time it was a waking experience, when, while motoring,
she saw him on the road in the form of an apparition, glaring with vicious
hostility. The incident occurred at a moment when she was in the act
of pulling off the road because her dog in the back of the car was
apparently in distress; it turned out that a leaking spray-can was emitting
fumes into the car. Why this should have provoked the apparition —
assuming the two data are related — is far from clear, nor why the
apparition should be the same one as she had seen in her hypnagogic
experience. But the salient characteristic is the sense of menace: it is
curious that this seemingly irrelevant emotion should emerge so
strikingly, particularly since in the later instance the apparition could
be interpreted as a well-intentioned warning of danger.

This air of malevolence does seem to be characteristic, for a minority
of people, of hypnagogic apparitions. Victoria Branden also tells of a
young boy who habitually saw 'scary faces', although 'he has learned
to reduce their menace by manipulating them, so to speak, into
something ridiculous'.

The letter quoted earlier indicates some of the difficulties of studying
this phenomenon: it is interesting that both this and Victoria Branden's
experience should have been followed by a different kind of
hallucinatory experience. The ability of entities to transfer, as it were,
from one mode of manifestation to another is an aspect we shall have
to take into account later. On the face of it, it is an argument in favour
of a single explanation for all entity sighting experiences. I think nobody
would doubt that the two types of phenomenon are closely related: at
the same time, it is clear that the hypnagogic experience has
characteristics which justify our studying it as a separate phenomenon.
Even if it should turn out to be classifiable as a subdivision of the
hallucination, it has its own distinct features, of which this extraordinary
parade of faces is the most notable.

If so, then the fact that it occurs only at the threshold of sleep must
be a crucial factor. What is so special about our state of mind at this
particular moment, which recurs at least twice daily throughout our

lives? Is the mind, as it makes the transition from one state to the other, in a particularly susceptible condition, like a soldier running from one patch of cover to the next, and so unusually open to external stimuli? Or is it that the experience is somehow physiologically necessary to us, and the function built into our system for reasons we can hardly even guess at?

One who did hazard a guess was Freud's disciple Herbert Silberer, who took a predictably psyche-oriented view. He considered that hypnagogic hallucinations are unique in the way they express abstract concepts in concrete form. The hypnagogic state is one of those in which the percipient is not in full command of his intellectual powers, and so his mind slips back into a non-rationalizing mode in which symbols are used instead of articulated thoughts.

I mention this simply to show how inadequate, for our purpose, most existing work on hypnagogic imagery has been. Silberer was able to relate some hypnagogic imagery to the percipient's current mental state with fair success, along psychoanalytical lines of course; but the scenes he describes could just as well be from sleeping or waking dreams: the unique features of hypnagogic image-seeing do not seem to concern him, he simply uses his findings to confirm those derived from other research. No doubt he could have found a psychoanalytical explanation for why Leaning's correspondent saw his cow and his duck, but he does not help us understand why this particular state produces this particular kind of experience.

We are going to find that, as in this case, there is often a substantial amount of overlap from one kind of sighting to another. Nevertheless we are also going to find that it is broadly true to say that if you are in a specific state, you are more likely to have one kind of experience than another: or to look at it the other way, if you have a certain kind of entity-seeing experience, the chances are that you are in one state rather than another. While it is far too early to start looking for reasons why this should be, clearly we shall eventually have to consider whether:

★ there is something about that particular state of mind which limits or dictates the type of experience which we can have while we are in it; or

★ whatever it is that causes or creates the experience, either gives us a particular kind of experience according to what state we happen to be in, or waits till we are in an appropriate state before giving us the experience.

We shall find that this is true of all the varieties of experience we shall be studying. Whatever it means, this characteristic feature can be used to support two points of view each of which is held by some people:

1   The pathological view, that the entity-sighting is the product, even the symptom, of a certain physiological state. For example, if you are suffering a delirious fever, you are liable to hallucinate: and if you hallucinate, that can be seen as an indication that you are in a delirious state, and therefore likely to be suffering from the kind of ailment of which this is a known symptom. This may or may not be true, but it still leaves open the question, why should experiences we undergo in one state take one form, but in another state take another?

2   The view that entity experiences, though superficially similar, in fact relate to a number of distinct phenomena, and it is an error which can lead only to confusion to suppose that psychedelic hallucinations, say, are related to mystical visions, and so on.

Again, this may turn out to be true. However, the fact that the correlation between states of mind and types of sighting is by no means total rather weakens these arguments: yes, they *are* distinct, but no, they are not *absolutely* distinct. In their ambiguity, hypnagogic phenomena perfectly demonstrate the entity enigma.

1.3 ENTITIES MANIFESTING AS HALLUCINATIONS

'Nearly every night a female, whose person is not unknown to me, comes close to me, throws herself on my chest, and presses me ⸱ ⸱ violently that I can scarcely breathe; if I cry out, she stifles me, and the more I endeavour to call out, the less am I able to do so.'

'There is nothing astonishing,' I replied, 'all this is only a phantom, and the effect of the imagination —'

'A phantom!' he cried, 'an effect of the imagination! I am telling you of what I have seen with my eyes, and touched with my hands. Even when I am awake, and with all my senses about me, I have seen her come and throw herself upon me . . .'[29]

The classic definition of a hallucination, originally formulated by Esquirol back in the 1830s, is very simple: it is the apparent perception of an object when no such object is present.

By 'perception' we mean that the percipient has no mere vague impression, but seems to see something which, however improbable, is extremely lifelike and often seen in such detail as to be indistinguishable from a real object. Here, for example, is a case reported by Abercrombie more than 150 years ago:

> The case of a gentleman has been communicated to me, who has been all his life affected by the appearance of spectral figures. To such an extent does this peculiarity exist, that, if he meets a friend in the street, he cannot at first satisfy himself whether he really sees the individual or a spectral figure. By close attention he can remark a difference between them, in the outline of the real figure being more distinctly defined than that of the spectral; but in general, he takes means for correcting his visual impression by touching the figure, or by listening to the sound of his footsteps . . . The gentleman is in the prime of life, of sound mind, in good health, and engaged in business.[1]

The philosopher C. D. Broad has observed: 'Up to the present, so far as I am aware, no one has managed to offer an intelligible concept, still less an imaginable schema, of the *modus operandi* of veridical hallucinations, which would enable a psychical researcher to infer what might be expected to happen in assignable circumstances.[13] It depends, of course, what we mean by hallucination. I am sure the reader is as reluctant as I am to get bogged down in questions of definition, so I do not intend to get more involved than I can help. But we must take note of some of the complexities.

For instance, an alternative use of the term restricts it to those cases where the percipient is *unaware* that what he is seeing has no basis in reality: when he *is* aware, then it should, some say, be called a *pseudo-hallucination*. Hilgard has described hallucinations as 'believed-in imagination', which rightly stresses the role of the percipient;·but just what is the percipient believing in, in the two cases I have cited? In practice, the distinction between hallucinations and pseudo-hallucinations is not very helpful, for there are many cases in which the only reason the percipient knows the hallucination to be unreal is because his commonsense, not his senses, tells him so — as in a case when a mountain climber encountered the bartender of New York's 21 Club, whom he knew to have been dead for five years, on a mountain peak [42j]. Had the climber's common sense not been in good working order, or had he not known the apparent to be dead, things might have turned out otherwise: as it happened, he knew at once that he was in the presence of a hallucination. The phenomenon would seem to be exactly the same in either case; it is only the percipient's response to it which is different.

Given our basic definition, every type of manifestation studied in this book could be a hallucination. For if we should come to the conclusion that there is no evidence that any of them is what it purports to be — i.e., an objective autonomous entity imposed or projected, visiting or intruding, from an external source — it would suggest that all these phenomena are simply various kinds of hallucination.

This may indeed prove to be the case, but at this stage we certainly do not have the right to assume so. Alternatively, we may find that the word 'object' in our definition is insufficiently precise. Does it take into account a case where the object the percipient thinks he is seeing is not there, but *something else is?* (I do not mean simply another physical object which is being mistaken or misinterpreted.) To take an extreme case, yet one which as we shall see has been seriously proposed, it is possible that there exist amorphous 'elementals' with the capacity to take on the likeness of, say, fairies, when circumstances are appropriate.

It is therefore convenient, at least for the time being, to continue to discuss the various phenomena as separate categories — apparitions, visions, *doppelgängers* and so on — rather than beg the question by labelling them all here and now as hallucinations. On the other hand, there are some kinds of entity sighting that are almost universally considered to be hallucinatory by nature. For example, I know of no pressure group which is lobbying for any alternative explanation for the visionary experiences of patients in delirium or fever, which may therefore be regarded as 'pure' hallucinations to which other varieties may be usefully compared. And yet even with this type of experience we may be begging the question.

Hallucinations are a fairly frequent side-effect or symptom of certain physical ailments. From the point of view of our study this has two consequences. It means that hallucinations are not only recognized by the medical profession as occurring, but have been positively studied by some of its members: as a result we know more about hallucinations than we do about most of the other forms of sighting experiences. But for the same reason, hallucinations have come to be generally associated with mental affliction, a pathological symptom indicative of a loss of mental balance or control. The inference has been that this is all we need to know about hallucinations: they are something that sometimes happens when we are ill, and when we get better again they stop happening. It does not seem to have occurred to most students of the subject that just as the content of our dreams may be more interesting than the fact that we dream at all, so it is not the fact that hallucination occurs, but that it takes the forms it does that is most interesting about it. The fact that a person hallucinates may indeed indicate that he is in an abnormal (though not necessarily pathological) mental state; but *what* he hallucinates may tell us a great deal more about him.

For let us be clear: *these hallucinations result not from the illness per se, but from the altered state of consciousness which is induced by the illness.* And though it *may* be that this state is brought about in no other way — that you have to have that particular illness to get into that particular state — there are no clear-cut ways of distinguishing one hallucinogenic state from another. It seems that there are several different ways by which people can get into a state where they are liable to start hallucinating; what they have in common is that all require the individual to quit his normal state of consciousness for some other.

To say that is not to say very much. Although certain specific states of consciousness have been isolated and labelled, we do not as yet know very much about them beyond a few physical factors like measurable brain-rhythms and so on. But clearly it is an observable fact that fevers lead to delusions and delirium; that drugs and alcohol, anaesthetics and opiates lead to a variety of visual experiences in which normal perception is modified; that alterations in breathing — underbreathing and overbreathing both alter the oxygen/carbon dioxide balance in the blood — often produce perceptual alterations; that fasting and sensory deprivation on the one hand, and stimulation and over-excitement on the other, can both distort our sense of reality; and that psychological stress, crisis situations whether real or fancied, can all prove favourable, if that is the word, to illusion.

Just how non-pathological and non-extreme a mental situation can be, and yet still generate hallucinations, is illustrated by this case:

> While engaged in the long-sustained and absorbing study and in the intensely imaginatively creative work of dramatic writing, I have not infrequently felt, seen, or heard one or more of my characters actually present and talking with me, and even at times critically considering my work. They were sometimes quite as vividly clear to me as if they had been truly objective,

but if I chose to reflect on their nature I was never deceived into thinking they were objective rather than subjective. I have therefore always believed that I was justified in regarding these 'visions' or whatever they might be called, not at all as abnormal phenomena, but as quasi-objective visualizations quite within the range of the normal imagination. [123]

To what extent can we compare hallucinations which seem to have physical causes to those which seem to be of purely mental origin? In a later section we shall see that there are indications that some kinds of difference exist between pathological and non-pathological hallucinations: but the study was made eighty years ago, and we have learned a lot since then. Without discounting the findings, which are certainly very interesting, we may feel some doubt as to the classification into two distinct modes, pathological or otherwise. The term 'psychedelic', used by researchers into the hallucinatory (and other) effects of drugs etc., [102] is equated with 'mind-manifesting', a phrase with obvious implications.

Yet again we have to say that, while this may well be an apt label, it is a question-begging one: in this case the question begged is whether the manifesting is taking place *from* or *via* the mind. Do psychedelic drugs stimulate the mind to forge hallucinations, or to modify sensory input; or do they put it into a passive state which facilitates the intrusion of external material?

The widespread nature of hallucination suggests that it is not so much a symptom of a specific state as a side-effect caused by a particular combination of psychological and physiological factors. It may be a form of communication or expression employed by the unconscious mind for some biological purpose: it may represent an act of opportunism on the part of some communicator, internal or external. Some cases seem to indicate such a purpose, while others seem quite arbitrary. Here are two contrasted cases:

On one occasion, when hurrying along a quiet street towards a railway station, Mr A suddenly checked his pace and said to his companion, 'Don't keep me so close to her, rather get in front. I was nearly on her skirt.' He was told that there was no one in front of him and no woman visible at the time on the same pavement. What Mr A saw as distinctly as anything he had ever seen in his life (if not more so) was the figure of a female walking closely in front of him, so that he could hardly step out without treading on her skirt. The skirt was of red cloth with groups of white lines crossing each other at frequent intervals, as in a tartan, and over this was a black silk jacket or short cloak. The dress was beautifully illuminated with sunlight and moved naturally in response to the motion of the figure, while the light silk jacket was occasionally lifted as if by the breeze. Mr A made a motion of putting the skirt out of his way with his umbrella, but of course, as he knew, there was nothing there. Having occasion to cross to the shady side

of the street, Mr A lost sight of the figure for a few moments, but it soon
reappeared right in front of him, with the sun still shining on it, but this time
she had changed her skirt and wore a rich silk tartan, the prevailing colour
of which was green. She was seen not very distinctly in the booking-office,
and finally disappeared for the day when the platform was reached. [148h]

Naturally there are several questions we would like to ask. Did the figure
make any sound? Would she have reacted if Mr A had indeed trodden
on her skirt? Would she have replied if he had spoken to her? (If only
he had leaned forward, tapped her on the shoulder, and asked her for
the time or a direction!) What we are left with is the apparition of an
unreal lady, unknown to the percipient, who behaved in a way quite
consistent with her environment, but was capable of actions — such
as the instantaneous change of costume — that were 'impossible' and
seemingly quite autonomous. They also seem to have been quite without
meaning or purpose, though that is something of which one can never
be finally sure.

Our second case is a first-person account of the experience of a Mr
Grunbaum, who was at the time undergoing hypnotic treatment (what
for, is not indicated) from a physician:

> I got aware of something moving and turning in front of and above my
> forehead. It took the shape of a disc of some four feet diameter. Inside that
> disc there was sitting a young lady. It was a beautiful creature with a very
> friendly charming face. She nodded her head very nicely towards me. I said
> 'Who are you?' She answered, 'I am your Self-Control.' (I had read in Dr
> Bramwell's book that the chief aim of every hypnotic treatment should be
> to develop the patient's self-control, but I had never thought of the idea that
> it meant to develop a young lady.)
>
> 'Just feel how real I am,' she said, and she stretched out her arm and hand
> towards me. So I gave her a pat on the fingers. I heard the sound it made
> and felt the touch. Then I noticed on that occasion something extraordinary:
> I felt in her hand just as well as my own hand. That is to say, that I felt just
> the same as when one is touching his own right hand with his left hand. My
> own hands were, however, not touching one another, but were lying on the
> woollen cover.
>
> Thereupon she began to make arrangements to step out of the disc. She
> put her foot outside. I still remember the beautifully decorated silk stocking.
> I could see every stitch of the silk. So I made directly up my mind that she
> had better stay in there, as I began to feel uneasy that there might something
> get wrong [sic] in my brains. She noticed my fear directly: I could see it on
> her face. So I returned to common consciousness and she disappeared. [148j]

Again, there are questions we would dearly love to ask. The phrase
about returning to 'common consciousness' indicates that the percipient
was in an *uncommon* state at the time and, moreover, that he knew
it: it would be immensely helpful to know a little more about that state.
But even as it is, the account is most revealing about the paradoxical
nature of the hallucination experience — the percipient observing

something even while he knows he 'shouldn't' be observing it. We note, too, the intensity of the experience: not only is it seen in vivid detail, but it is very striking in its form of presentation. Whatever its origin, it was clearly designed to make a very strong impression on the percipient. The remark 'I am your Self-Control' suggests some kind of external expression of a mental process, the personification of an abstract idea currently preoccupying his thoughts. Mr Grunbaum himself 'never seriously supposed that anything beyond his own mind was at work', the report informs us, and the immediate 'explanation', that the vision was simply a projection from his unconscious mind, may well be basically a sound one. The similarity to the dream-experience suggests that the same production team is responsible for both types of phenomenon: but it is equally clear that the hallucination-experience is something distinct from the dream-experience, not only because it occurs in the waking and conscious state (though what kind of 'conscious' remains debatable) but also because it is presented as a logical, realistic incident within the actual, immediate circumstances in which the percipient finds himself. In dreams, we are transported to whatever location the producer has chosen, whether real or imagined; but in a hallucination the experience generally occurs 'on location' — wherever the percipient happens to be. Moreover, while the dream is 'out of time', immensely complex narratives being encompassed in minutes if not in seconds, this is not true of the hallucination, which appears to occupy 'real time', though with off-stage modifications such as the lady's change of clothes in Mr A's case.

Though we may suppose the percipient of a hallucination experience to be in a specific state of mind, it is a quite different state from that in which dreams or hypnagogic visions — to take simply those experiences we have looked at so far — are experienced. For one thing, while only a very few dreamers are able to exercise any kind of control over their dreams — and when they can, it is liable to be limited to terminating them rather than directing their course — hallucination percipients are, in some cases at any rate, able to modify the experience by their own participation, as Mr Grunbaum did. Some, such as the gentleman cited by Abercrombie earlier in this chapter, are actually able to initiate a hallucination voluntarily: but, curiously, he was not able to terminate the proceedings at will, but had to wait until the entity chose to leave of its own accord.

Most hallucination percipients are taken by surprise by their visitors, however, and the circumstances are dictated by the entity — who frequently appears to act quite autonomously. In most cases, though, the entity adapts to the environment and reacts to the percipient: it gives every impression of being aware of the percipient's thoughts and actions.

The hallucination experience is, as we shall see, crucial to our inquiry.

If, as seems probable, it is initiated by the unconscious mind and presented to the conscious mind via the normal perceptive process so as to give the mind the illusion of reality, then it is evidence of a very remarkable process. In considering dreams we saw how they require us to recognize the existence of an autonomous 'producer' who is able to call on immense resources both of creativity and material. Hypnagogic experiences extended the range of the phenomena and provided further evidence of the independence of the creative force from our conscious minds. And now hallucinations suggest that such productions are not limited to those times when our conscious minds are closed down or resting, but that the producer can 'put on a show' capable of convincing the conscious waking mind of its reality. In the case of Mr A, for instance, would he ever have discovered that the lady was an illusion if he hadn't had a companion with him, or if she had been further away so that the question of her reality was never raised?

We cannot at this stage say whether dreams, hypnagogic visions and hallucinations all emanate from the same source, or have any common purpose. We have still to account for the fact that there are marked differences between these kinds of experience, and we cannot yet eliminate the possibility of an external agent. One thing, however, can usefully be emphasized at this stage: we have no reason to suppose that the act of hallucination is anything but a fairly common mental process. The variety of circumstances in which hallucinations are generated makes it clear that we are dealing with a base activity which is not in itself an indication of any anomalous condition. As W. F. Prince wrote in 1930:

> No species or instance of hallucinations can in itself be evidence for the supernormal. The evidence, if any, is to be found in the relation of the hallucination to something else, with no discoverable causal link connecting them. [123]

In the ensuing sections, then, we shall be looking at cases that seem to link the act of hallucination with 'something else'. It is this additional element, which so often appears to take the form of some*body* else, that has encouraged many to believe that some kinds of hallucination-experience are anomalous to the point of requiring a paranormal explanation, and many others to believe that certain experiences, though seemingly hallucinatory in nature, are not hallucinations at all but something much more rich and strange.

1.4 DOPPELGÄNGERS AND ASTRAL DOUBLES

I knew a very intelligent and amiable man, who had the power of placing before his own eyes himself, and often laughed heartily at his double, who always seemed to laugh in turn. This was long the subject of amusement and joke, but the ultimate result was lamentable. He became gradually convinced that he was haunted by his self. This other self would argue with him pertinaciously, and, to his great mortification, sometimes refute him, which, as he was very proud of his logical powers, humiliated him exceedingly. At length, worn out by the annoyance, he deliberately resolved not to enter on another year of existence — paid all his debts — waited, pistol in hand, the night of the 31st of December, and as the clock struck twelve, fired it into his mouth.[29]

'The mystery of the human double', to quote the title of a classic book on the subject, is one which makes nonsense of any formal categories of entity phenomenon. We can distinguish at least five kinds of experience in which recognizable facsimiles of living persons are seen apart from their physical bodies:

1 'Phantasms of the living', as the SPR termed them in 1886.[61] In these cases the apparition of someone 'known to be living at the time' is seen where that person certainly is not present.

2 Doubles of living people seen in close proximity to the original, if not in immediate juxtaposition then at least so close that it is often possible for both to be seen at once.

3 Bi-location of a person who seems to be conducting himself in a normal fashion in two places at once. This is specially reported of religious figures, and is treated by writers on the physical phenomena of mysticism as one of many miraculous feats recorded of saints and others known for such phenomena.

4 The deliberate or at least wish-associated projection of one's double, usually on an experimental basis, as when an agent causes his double to manifest in the distant presence of a friend, sometimes with and sometimes without prior arrangement.

5 The 'doppelgänger' phenomenon in which the percipient sees his own image.

On the face of it, it seems that the first category is a kind of ghost, and will be discussed in the section devoted to apparitions: category 3 may be the same, though there are some differences, particularly in that the original and the double both seem to behave normally at the same time. Category 4 will be more fully discussed in Part Two of our inquiry, when we consider experimental cases. It is only with the last category, the doppelgänger proper, that we are here concerned, and with the second category inasmuch as it seems to be at least as closely related to the doppelgänger as to the ghost.

But simply saying that, makes it only too evident how presumptuous it is to draw hard-and-fast distinctions between these types of case. The doppelgänger has all the indications of an hallucination, and is generally so treated, if for no other reason than that it is generally perceived by nobody besides the percipient himself. But though this is not true of our second category, the common features of the two categories justify us in treating them together: we then see how closely they resemble cases in category 4 . . . and so it goes on. So, before we concentrate on the doppelgänger *per se*, we should give some thought to some of the general principles involved.

Many people believe that we all possess a second body, often termed 'the astral body', which is normally coincident with our physical body but can under certain circumstances separate from it. Attempts have been made to establish the existence of the astral body on a scientific basis [3, 80], but without convincing success. This has not prevented very elaborate models being proposed, some of which hypothesize more than one secondary body. (Robert Crookall gives the most explicit and convincing account of this. [25]) One of the attractions of the astral body hypothesis is that it accounts neatly for out-of-the-body experiences: it is suggested that in such cases the astral body separates from the physical body and goes wandering off on its own. Which, at such times, is the *real* self, becomes a meaningless question, as academic as when a similar question is asked with regard to the multiple personality cases that are also germane to our inquiry and which we shall look at later.

Not so academic, though, is the question of where the 'seat of consciousness' is located when the astral body separates from the physical. In the archetypal instances in which, typically during a serious medical operation, the patient seems to be floating in mid-air looking down at his body, it certainly seems as though the seat of consciousness has been transferred to the non-physical body, whether we think of this as 'astral' or otherwise. This has been a severe stumbling-block to those who favour any alternative explanation, and though it has been proposed that the whole experience is an imaginary one consisting of a 'dramatization' of data obtained by extra-sensory perception of the 'travelling clairvoyance' kind, this is not only pure speculation but also rather cumbersome. It is easy to see why the astral body explanation is preferred.

The evidence for a real process of separation, as opposed to an imaginary one of the kind just mentioned, is supported by subjective testimony. The 'astral voyager' in many cases reports on the effort of separation from the physical body, and subsequently on his reluctance to return to it. True, all this could be part of the ingenious scenario prepared by the subconscious mind; but if so, it is curious that so many people's subconscious minds devise the same scenario.

In considering the doppelgänger and other cases of bi-location, then, we have for once a very specific model offering itself as one possible explanation. It is not, as we shall see, the only model available: but if it were to be established that the astral body is indeed a reality, it would have to be anyone's first choice to explain what is seen when a doppelgänger or a phantasm of the living is reported.

Let us start with a typical and uncomplicated case:

> One day in 1958, Harold C of Chicago came home after his day's work suffering from an attack of migraine. He sat down for his dinner and saw, sitting opposite him, an exact replica of himself, who copied every movement he made. It stayed throughout the meal and then vanished. There was no obvious purpose in this episode. He had had other similar experiences, always associated with his migraine attacks. [42f]

Autoscopy, as it is technically known, is sufficiently common to be recognized by doctors, who have proffered physical explanations for it. The fact that it is so often associated with migraine and epilepsy points clearly towards some physical explanation in terms of brain function, and one school of thought accounts for it as the result of some irritating process in the brain, specifically the parieto-temporal-occipital area. However, such an account, even if valid, does no more than outline the mechanics of the process: it does not explain why Harold C's migraines produced this effect while other people's migraines do not. Nor, to go deeper still, does it demonstrate that the migraine caused the double to appear: can we be sure that Harold's migraine was not a preliminary warning that he was about to have a doppelgänger experience, just as some other witnesses to paranormal phenomena claim to have had intuitive warnings beforehand?

Harold C's sighting seemed to serve no purpose, unless his astral body, unwilling to submit to the pain of the migraine, decided to separate itself from the physical body until the migraine had passed away . . . There are other cases, though, where a psychological motivation is clearly indicated:

> One day in 1962 Mr John S of Boston was sitting at home listening to classical music on records when he looked up to see a duplicate of himself leading an orchestra playing the music he was listening to. [42f]

We can hardly doubt that the hallucination in this instance represented a wish on John's part, reminiscent of Oscar Levant's dream-sequence in *An American in Paris* in which the unsuccessful musician sees himself not only conducting the orchestra but also playing all the instruments. John seems likely to have been in a relaxed, contemplative mood induced by the music, favouring such a manifestation. But why, in that case, don't more of us have such experiences?

The following case, too, seems to have arisen from the percipient's state of mind rather than from any physical condition:

> On the 15th of March, 1978, at 10 o'clock at night, I saw an apparition of myself. I was alone: a friend who had been with me had left about half an hour before, and I had been working at the sewing machine. One of the children was sleeping restlessly; I took the lamp to see if anything was wrong. As I drew back the curtain which shut off the bedroom, I saw two paces from me the image of myself stooping over the end of the bed, in a dress which I had not been wearing for some time: the figure was turned three-quarters away from me, the attitude expressed deep grief . . . I was neither specially sad nor specially excited that evening, and had been thinking about quite ordinary things. Three months before, I had lost one of my children. It has just occurred to me while writing this, that after death my child was laid across the foot of my bed, and I may have stood in that attitude then. The dress, too, was the one I was wearing at the time. [116]

Various explanations offer themselves, including that the room was haunted by her previous self; but a psychological explanation seems the most plausible, such as that the entity was a projection from her subconscious mind, which was less happy than she was trying outwardly to be. One could add that working at a sewing machine is ideal for inducing a state of detachment which could lead to psychological dissociation and thus make things easier for the projection to occur. But the question arises yet again, why did she have this experience when so many in similar situations do not?

An additional and very awkward dimension is introduced in the following case:

> About 1951, the percipient was ill with some kind of virus infection for which her doctor had prescribed some drugs which seemed to make her worse, not better. 'I was lying in bed, knowing I was there quite comfortably, when I suddenly saw myself sitting on my bedside chair, dressed in a frock discarded quite a year before. I did not speak at all, but myself in the chair told me that if I wished to recover I should stop taking the tablets at once or they might finish me off. It passed through my mind, Why bother if they did finish me off? However, my chair self said that was a stupid thing to do, and finally persuaded me to stop them immediately and tell the doctor, and then she disappeared.' She stopped the tablets, and immediately began to get better. [58]

Timely warnings from entities is a motif which runs throughout our inquiry; we shall find them uttered by secondary personalities and by spirits of the dead, by religious visions and by 'guardian' figures, by alleged extraterrestrial visitors and, as here, by 'astral doubles'.

Whatever the explanation for this diversity of kindly interest, we have to recognize that it represents a degree of 'higher knowledge' on the warner's part: someone or something knew that the tablets in the foregoing instance were doing harm to the patient, and was concerned to warn her. We could conjecture that it was the patient's own subconscious mind dramatizing the information in this way in order to give force to the warning; or we could suppose that some external agency chose to appear in this guise. But whatever the explanation, it seems beyond question that, in this case at least, the doppelgänger was a purposeful manifestation created for the specific occasion.

But not all doppelgängers are so well disposed:

A 27-year-old Okinawa woman was admitted to an American hospital covered with blood from a cut in her tongue, and her face was swollen and congested from what seemed like a strangling attempt. She said that, while shopping, she had met a woman looking exactly like herself, dressed in the same clothes, but with a 'sharp, unkind expression'. The witness was carrying two parcels; suddenly she missed one, but when she got home she found it waiting for her.

Next day she felt ill; at midday there was a knock at her door. It was the same woman, who spoke to her in poor Japanese, saying 'Let us go inside'. The percipient left her small baby with neighbours, then went into the apartment with the apparition, who proceeded to attack her, trying to strangle her and to cut her tongue with a pair of scissors. She passed out; when her husband came home he found her lying on the floor and rushed her to hospital. [54]

It was evident that the wounds were self-inflicted, and equally evident that the attitude and the actions of the double reflected her own fears and preoccupations: the medical adviser on the case, noting that she was an alien living in a strange country with demanding parents-in-law, and with a six-month child, could not do other than attribute the attack to an exteriorization of her personal problems; I can see no reason to differ from this diagnosis. But once again, why did the manifestation take this particular form?

A percipient who underwent a somewhat similar experience was able to add her own comments. Victoria Branden, the Canadian author whose hypnagogic experience we have already noted, tells of her doppelgänger experience:

I drove round a curve in the highway, and found a huge truck coming right at me, on the wrong side of the road. It is difficult to describe what happened next. I didn't black out, but there was a sort of suspension of consciousness, so that instead of seeing a truck, I saw beside me in the passenger seat — myself. It was dressed in a starched blue and white cotton dress (as I was); its hands were folded passively and its head dropped forward on its chest . . . What really scared me was that the figure beside me, for all its passivity,

seemed indescribably malign and threatening. I imagined that the meekly
bowed head had concealed an evil smile. [11]

Branden reasonably relates the vision to her mental state at this critical
moment, but assumes it to have been a projection of her subconscious
mind: if so, however, it is difficult to see what purpose it served — it
did not help at all in the immediate situation, but exacerbated its
traumatic effect. If, as Victoria Branden surmises, it was a symbol of
her own guilty feelings (unjustified, as she recognizes) about a recent
motor accident in her family, it seems a curiously obscure form of
expression.

So far, in this chapter, we have concentrated on true doppelgänger
cases, in which the percipient sees himself. We should also note,
however, some cases which seem related, though with the mental
reservation that the relationship may be illusory. Here, briefly, is one
of the best-known double cases in the literature:

> From 1845 to 1846 Emelie Sagée, aged thirty-two, was a teacher at a girls'
> school in Livonia, now part of the USSR. She was French by birth, and all
> comments on her attest to her good character, pleasant disposition and
> consequent popularity. Unfortunately, she caused much disturbance in the
> school by being repeatedly seen simultaneously in two places. Sometimes
> it was a case of her double standing right beside her, standing at the blackboard
> in the classroom, say, the double precisely mimicking the actions of the real
> teacher: sometimes the two were separated by a considerable distance, as
> once when the real teacher was outside in the garden, seen through the
> window, while her double sat in a chair at the head of a table in the classroom.
> These incidents were seen by a great number of percipients — thirteen saw
> the two Emelies at the blackboard — and forty-two — the entire school —
> saw her simultaneously indoors and in the garden.
>
> It was noted that when this separation was occurring, the double seemed
> to draw strength and life from the physical Emelie, who appeared drowsy
> and exhausted until the double vanished, at which time she would revive.
> A significant clue is afforded by Emelie's own statement, on the occasion
> of the indoors/outdoors sighting, that while in the garden she had been
> thinking that someone ought really to be in the classroom supervising the
> girls in the absence of the headmistress.
>
> On that occasion two of the girls, who because the phenomenon had
> occurred so frequently had to some extent lost their fear of it, dared to touch
> the figure. They averred that they did feel a slight resistance, which they
> likened to that which a fabric of fine muslin or crape would offer to the touch.
> One of the two then passed close in front of the armchair, and actually *through*
> a portion of the figure'.
>
> However, though some of the pupils may have lost their fear, the stories
> began to affect the fortunes of the school, and Emelie was asked to leave.
> She then revealed that this was the nineteenth time she had lost a job as the
> result of her curious manifestations. [115]

Though the story as we have it is very circumstantial, it rests on the
testimony of a single witness, one of Emelie's pupils, recorded fourteen

years after the event. Attempts to follow up so interesting a case have been made, and they have in part confirmed its truth, but that is far from absolute reliability. It has enough in common with other cases, however, to justify us in giving it provisional acceptance. So far as is known, Emelie did not see her own double, but this does not emerge clearly from the report and is probably the first question we would wish to ask if we had the chance.

A case which seems half-way between the Sagée episode and true doppelgänger cases is the following:

A Mrs Milman, wife of an official in the Houses of Parliament, London, stated that for years she had been 'afflicted with another self, that people meet in places where I am not. The other day a friend took leave of me in the work-room. Scarcely had he stepped out at the door, when he found me again on the landing of the stairway. Dumbfounded, he shrank aside to let me pass. I had not stirred.

'I have never seen my double, but I have heard it. One evening, just as I had come into my room, I heard a cracking sound and went out upon the landing. All the doors that I had just closed were open. I went back precipitately and rang for the servants . . . The House-steward was most surprised to find me in my room, since he had just seen me, he said, opening the door of a hall on the ground floor.'[43]

Strictly speaking, these are not doppelgänger cases: they are phantasms of the living, and whatever explanation we find for ghosts and other apparitions, is likely to apply to them. But are we justified in drawing a hard line between these cases and the true doppelgängers? Was the second Mrs Milman, who was seen opening doors by the servant, different in kind from the second mother seen by herself leaning over the child's bed?

On the face of it, there *is* justification for making a distinction. True doppelgängers are not usually, if ever, seen by anyone other than their originals. This suggests that they represent a personal response to a specific situation. From a medical point of view, there is a clear tie-up in some cases with certain physical states: but the nature of migraines and epilepsy is insufficiently known and we cannot discount a psychogenic origin for them, in which case both physical state and vision are alike effects with a common psychological cause. Thus, in our first case, Harold C's migraine and his vision may *both* have been caused by some personal preoccupation, on either a conscious or subconscious level.

If so, then the doppelgänger may be simply a way chosen by the subconscious to resolve the situation, either by provoking a crisis, or by uttering a warning. But this hardly seems to cover all our cases: for instance, it does not appear to fit Victoria Branden's experience or that of the Okinawan lady. In such cases it almost seems as though a secondary personality of the percipient may be involved, and I suppose it is plausible that, when there is sufficient fragmentation of the personality — if the Okinawan lady was in the process of a mental breakdown, if Victoria Branden was momentarily 'shattered' by her near-accident — we could suppose a malevolent aspect of the personality assuming this appearance. Or, on another hypothesis, we could suppose that a mischievous spirit did likewise . . .

Clearly, there are many different ways of looking at the doppelgänger phenomenon, and it would be unwise of us to make any premature judgments until we have studied some other types of manifestation. What is certain, though, is that — like each of our categories — the doppelgänger is a very complex phenomenon, which seems to grow more complex still the more we study it. We have already seen that a physiological explanation is inadequate on its own: but we further see that even a psychological one is insufficient to account for dimensions which demand a parapsychological explanation. Consider this famous case: that of the German writer Goethe:

> I was riding on the footpath towards Drusenheim, and there one of the strangest presentiments occurred to me. I saw myself on the same road on horseback, but in clothes such as I had never worn . . . As soon as I had aroused myself from this daydream the vision disappeared. Strange, however, it is that eight years later I found myself on the identical spot, intending to visit Frederika once more, and in the same clothes which I had seen in my vision, and which I now wore, not from choice but by accident. [53]

Not only is the double capable of knowledge that is hidden from our conscious thoughts, as in the case of the tablets cited above (page 62), but it is also apparently capable of precognition. Indeed, it would seem that the doppelgänger is liberated from the normal bounds of time and place, just like the producer of our dreams. If it is a parasite, or a simple hallucination generated by our disordered minds, it seems curious that it should be invested with this enhanced access to information. But there is little evidence that it is anything more; in which case we have, not for the first nor the last time, to face the fact that our subconscious producer has privileged access to knowledge — or freedom from restrictions on such knowledge — denied to our conscious selves.

1.5 COMPANIONS AND COUNSELLORS

When she was about four years old, the psychic Eileen Garrett first met 'The Children':

> There were two little girls and a boy. I first saw them framed in the doorway; they were strange to me, as were all children; I looked at them intently and longed to play with them, but I was not allowed to mix with other children . . . Out-of-doors next day, I saw them again. I joined them, and after that they came to see me daily. Sometimes they stayed all day, sometimes but a little while, but no day passed of which they were not a part. Until I was thirteen, they remained in touch with me, coming and going . . . My aunt and uncle ridiculed the whole idea: 'Now do be quiet about those children of yours. You are making it all up; there are no such children, and God will surely punish you for telling such lies.'
>   I never doubted the reality of 'My Children'. I touched them and found that they were soft and warm. There was one way in which they differed from other people. I saw the form of ordinary humans surrounded by a nimbus of light, but the form of 'The Children' consisted entirely of this light. [50]

To a considerable degree Eileen Garrett's companions had lives and natures of their own. Though they shared in most of her games, they did not like climbing trees, as she did, and they seemed unhappy with water. They showed her many things, explained things, taught her things. Yet all this was accomplished without the use of words.

The class of beings often referred to as 'childhood companions' is apt to be cited as an instance of the operation of 'eidetic imagery', a phenomenon which has been rather unsatisfactorily defined as 'a type of vivid imagery which is, as it were, projected into the external world, and not merely "in one's head"; a half-way house to hallucination; also the ability or disposition to project images, frequently characteristic of children'. [33]

Certainly the phenomenon is chiefly associated with children. It is of course common practice among children to 'dream up' imaginary playmates, who share in their games, act as confidantes, and give psychological support and comfort. Probably most children, at a certain stage in their development, can be seen wandering about the garden on their own, deep in communication with an unseen companion. But Eileen Garrett's experience seems to have been much more profound than this. So, while the obvious supposition must be that she herself 'created' the children to fill a psychological need, there are several puzzling aspects to the case. Why, one wonders, did she go to the trouble of inventing not one but three children? Why, if they were simply figments of her imagination, did they not share her fondness for climbing trees? We could invent answers to those questions, such as that the three were modelled on a book she had read, and that their disinclination to climb trees reflected her own subconscious fears of doing so: but whence came the information and explanations they were able to give her? To say that Eileen Garrett was in later life a celebrated psychic, and no doubt had the seeds of this capability even at the age of four, explains nothing except insofar as we may credit her with an above-average accessibility to psychic influences. But if the 'Children' were generated inside herself, how does the question of psychic influences touch the matter?

She was often asked whether her 'Children' resembled any that she had known in real life: the answer was no. This does not necessarily mean that they had no real-life counterparts, it simply leaves the question open. They could, for example, have been the ghosts of children who had formerly lived in that house or nearby — a favourite theme among writers of ghost stories. But how does Eileen Garrett's experience relate to a case I encountered personally, in which a lady told me how, throughout her rather difficult childhood, much of which was spent with nurses and guardians, she had received comfort and support from her twin sister — who had died at the moment of birth?

The situation had been such that only one of the twin babies could be saved, and it chanced to be my informant and not the other. She had felt, ever since she learned of the circumstances, a sense of guilt that it had been she who had been preserved, and this certainly provided a strong emotional force that no doubt played its part in the matter; we could surmise that it translated into a sense that she owed it to her dead sister, as it were, to share her life with her, and so 'brought her into being' with that intention. But having done so, the sister seemed to take on an autonomous life of her own: she was a continual support and solace to the living sister, and seemingly more knowledgeable. She enabled my informant to have precognitive knowledge of certain events, and on one occasion, taken to Chichester for the first time, the living sister — then I think aged about twelve — knew her way about the town, crediting the knowledge to her twin. Talking to her, it was hard even for the rest of us to avoid feeling that the dead twin was a real active personality, and there is no doubt that she was so to her living sister. 'Wish-fulfilling fantasy' seems a feeble description of so intensely felt a presence.

The following case is even more complex. An anonymous Russian lady wrote to the *Journal* of the SPR as follows. Her account commences before her sixth birthday:

My earliest recollections were of a strange nature: every night, after I had been put to bed, I saw the door open and a strange woman come in, who glided up to my bedside and sat down on a chair, bending over to look at me. She had masses of grey hair, done up in a fashion quite unlike to the people I was accustomed to see, as was also her dress . . .

At first this apparition threw me into convulsions of terror, which were not lessened by any caresses bestowed upon me, and the assurance of my mother and nurse that 'there was no such woman'. I only resented such assurances, for how could I not believe what I saw, palpable, alive, as much so as were the people surrounding me. Everything was tried to break the spell; I was put into a different room, my mother sat upon the chair by my bed trying to soothe me. It was no use! Just as before, night after night, I saw the figure glide into the room, and the chair being occupied, she stood at the other side of the bed, or at the foot end, watching me . . .

Little by little I grew accustomed to her coming: I even watched for it and grew to like it, for my child's instinct told me that this woman was my friend, and had some kindly purpose in watching by my bed every night . . . She remained with me for hours, and sometimes put a cool, slender hand on my head . . . She never spoke to me, nor I to her, and this seemed natural to me: I felt that we understood one another . . . This apparition visited me daily till I was nearly ten years old. Sometimes she also came by day and watched me playing . . . After I had reached the age of ten, my apparition ceased to come to me regularly — her visits became less and less frequent, and at last stopped altogether. [147c]

Such a case as this is strongly suggestive of a haunting, but at the same time the clear personal link with the percipient justifies our classing it as a kind of 'companion' case: though no counselling took place, a relationship of a sort developed between entity and percipient, and the nightly recurrence of the visits over some four years is to be compared with Eileen Garrett's daily experience. Significant, too, is the fact that both visitations gradually ceased as the percipient entered her teens.

However, there was an important additional dimension to the Russian lady's case. As she grew older, she became aware of many hints of a link between her and the court of Marie Antoinette in eighteenth-century France. In her lessons and her reading she developed an intense interest in that period; when she visited Versailles she displayed that ability to know her way about, to recognize details, to explain the function of the rooms and so on, which is generally called *déjà vu*, though in this case it does seem as though knowledge had been somehow acquired on a deeper level. Her own inference was that her childhood companion *was* Marie Antoinette, implausible as the idea might sound:

Of course every child, and even most grown-up people, may have a special sympathy for some figure in history, but mine was more than an ordinary sympathy, it was a cult, an obsession. I spent hours at the South Kensington Museum gazing at Marie Antoinette's bust, examining her toilet-table with its little rouge pots etc. . . . . I very frequently dreamed of her, and though my dreams generally were very disconnected and intangible, whenever I dreamt of her it was in a most logical sequence, and, contrary to most dreams, I remembered every detail in the morning. These dreams represented a routine of daily life at some palace — always the same palace — but one I had never seen in reality.

As may be expected, when she did get to visit Versailles, there was, besides the familiarity with the lay-out of the palace, a great emotional response. Of course, given our knowledge of the events related to Versailles, few of us are likely to visit it without strong feelings, and this may well be the reason why it has been associated with many memorable 'psychic' episodes; nevertheless, there seems no reason to doubt that a special affinity was operating in this case. It is tempting to interpret it in terms of past lives, that perhaps she had indeed been at Marie Antoinette's court in a previous incarnation and this gave her the necessary link with the past in her new life. We know enough about such reports to be on our guard against taking them at their face value: but such an explanation has a greater plausibility than that the spirit of Marie Antoinette came to visit a little girl who, at the age of five, presumably had no notion who she was and certainly lacked the knowledge to create her own do-it-yourself Queen of France!

In the following case the companion unquestionably had a recognized real-life counterpart: the percipient was an eighteen-year-old girl whose adored grandfather, educated and intelligent, had been in the habit of discussing her school problems with her and helping her with her studies, which he much enjoyed doing:

> When he died I was about eighteen, and I was desolate. About six months after his death, I had another big problem to resolve, and no one to talk it over with. During that week once at about 3 a.m. my grandfather came through the outside wall of my bedroom, in a luminous circle of grey light, his head and shoulders clearly visible, and talked to me. Next day I found I could understand my problem, and could proceed with my work. My 'grandfather' came again some months later when I again needed his help — but he has never come again since. [58]

Though at first sight it would seem that the 'grandfather' returned in response to his granddaughter's need, we should also take into account the fact that he had been very fond of helping her. This could explain why he came to help her while so many other children in need of assistance have to go without. The implication is that it takes two to forge such a relationship, that it results from the combined wishes of percipient and entity.

Such a hypothesis would also enable us to get round some of the puzzling aspects of the Russian lady's case, and explain how the little girl came to be visited by an entity who — whether or not she was the spirit of Marie Antoinette — was surely in a form the girl could have imagined for herself. If we suppose that the girl was unconsciously in need of companionship, perhaps her need was picked up by the spirit of the dead lady, who may have had a need of her own that corresponded, and responded, to it. The rest — the passion for everything associated with the earthly existence of Marie Antoinette — would follow naturally enough.

Some such two-way process, whereby an entity manifests in response to a need of the percipient's, also seems to come into operation when, in a moment of danger or crisis, a witness reports feeling that he has someone with him, comforting and protecting, sometimes even giving tangible help and advice. We shall be looking later at some instances of this kind of experience: but we should recognize here that, though there is a shared element with childhood companions — the comfort and advice — there is also a major difference. This is that the one-off cases seem unquestionably to occur as the direct response to a specific crisis-situation in which the percipient finds himself: the companion cases, on the other hand, describe a continuing relationship to which the specific acts of assistance seem to be only incidental.

Possibly, though, the one-off cases are not as unique as they appear to be; they may be momentary manifestations on the conscious level of a relationship that is otherwise maintained on an unconscious level: if so, the underlying relationship could conceivably be much the same in both cases.

What form would such a relationship 'really' take? It is tempting to speculate that it could be something along the lines of a 'guardian angel', a benevolent being assigned to each of us on a permanent basis, who normally works in the background like a discreet family servant, but who sometimes emerges in response to specific needs, whether momentary — as in a physical emergency — or continuous, as in the case of childhood companions. The circumstances of those needs, we might further speculate, would dictate the form in which the 'angel' manifests: to a child needing companionship he would appear either as another child, or as a mother-surrogate; in crisis situations he would perhaps appear as an authority figure, in conformity with whatever cultural beliefs the percipient held — an angel to a Christian, a bodhisattva to a Theosophist, and so on. However, attractive though such a hypothesis may be, it must be admitted that it does not seem to help much with such cases as the 'Marie Antoinette' case just cited.

There is, however, a category of case which seems to span companion-cases and crisis-cases alike. Edith Foltz-Stearns was a woman pilot who, from the time she got her licence in 1928, was convinced that she was watched over by protecting 'companions':

> I never fly alone. Some 'presence' sits beside me, my 'co-pilot' as I have come to think of it. In times of great danger some unseen hand actually takes the controls and guides me to safety. [42i]

This suggests an on-going relationship — except for one disconcerting feature: the 'co-pilot' is not the same person every time. In 1932 she was warned not to make a dangerous forced landing thanks to 'the voice of an old school chum who had died when she was fourteen': while ferrying a plane during World War Two it was her dead father's voice that saved her from flying into a mountain.

Here we have the option of concluding that Edith had a number of helpful well-wishers who intervened from another plane, or that the helpers were not who they purported to be, but the same guardian in a number of disguises. One can understand that if the warnings emanated from her subconscious mind, they might appear to come from specific individuals; but it is certainly curious that, while feeling this to be a continuing relationship, the relationship should not be always with the same entity.

Another category of cases that seem as though they might relate to the 'companions' are the 'extra man' cases sometimes reported, particularly in circumstances of extreme physical demand. In 1916, during World War One, Commander Stoker and two companions escaped from a prison in Turkey. During their five hundred kilometre escape journey to the Mediterranean they were conscious of being accompanied by a 'fourth man'. None of the three mentioned it at the time, and it was only when they compared notes afterwards that they realized that it had been an experience they had all shared.[42g]

By an odd coincidence — if it is nothing more — another of the best-known instances of this phenomenon dates from the very same year. Sir Ernest Shackleton, the Antarctic explorer, was making a demanding crossing of the mountains of South Georgia, during which both he and his companion felt there was an extra person with them.[42g] The climber Frank Smythe has provided a further example, which occurred during his attempt to climb Everest in 1933, at a time when his companion had had to turn back:

> All the time that I was climbing alone, I had the feeling that there was someone with me. I felt also that were I to slip I should be held up and supported as though I had a companion above me with a rope . . . When I reached the ledge I felt I ought to eat something in order to keep up my strength. All I had brought with me was a slab of Kendal mint cake. This I took out of my pocket and, carefully dividing it into two halves, turned round with one half in my hand to offer to my 'companion'.[146]

And so we find that a phenomenon which at first seemed to be easily attributable to 'eidetic imagery', and readily explained as the expression of childhood loneliness, turns out to have several disconcerting facets, some of which seem to point suggestively towards the participation of some external agency.

Just how complex companion cases can be is demonstrated by this recent example.[147f] In a house in Bournemouth, England, troubled with poltergeist activity, some spiritualist seances were held in August 1981, in the hope of learning the cause of the disturbances. In the course of one of these, the medium was controlled by — among others — a little boy who said his name was Ian and who claimed that he often played with Bradley, the eight-year-old retarded foster-son of the family. And indeed Bradley, though he had a real-life friend named Ian whom he played with, stated that when he was on his own he played with another friend named Ian.

The seance-Ian returned at a later seance, but this time he described himself as a young man. Asked how he could be a little boy in the morning and a young man in the evening, he answered that time ran differently where he was and that 'you are what you think you are'. He identified himself as a 22-year-old man who had died in the locality; he said he came back to play with Bradley because he too had been an unhappy backward child. (It was confirmed that a young drug addict named Ian had indeed died in the area some two years previously.)

Our options are:

1   The seance-Ian emanated from the mind of the medium, who we may suppose to have been well acquainted with the facts of the case. If not, he could conceivably have obtained access to them via ESP.

2   The seance-Ian was 'picked up' by the medium from the mind of Bradley, and dramatized by the medium into a 'real' spirit personality. We would suppose that Bradley had created his companion Ian from stories he had heard relating to the death of the young man.

3   The surviving spirit of the dead young man Ian does in truth return to Earth to play with Bradley, just as both parties claim.

In many ways this third alternative is the most attractive. Though it requires us to make the boldest assumptions, it has a satisfying logic to it. If it should turn out to be the fact, it would encourage us to believe that many if not all of our 'companions' are discarnate entities, rather than creations of the percipients' imaginations. But so long as a plausible psychological model is available, we must reserve our judgment.

Companion cases have similarities to several other categories of entity-sighting. Even the handful of cases we have looked at suggests comparison with hallucinations, with ghosts and hauntings, with spirit guides, with regression and reincarnation cases; and we have the case of Glenda, with which I opened this whole inquiry, to suggest a link even with extraterrestrial beings. At the same time, there are manifest differences between the cases we have cited. Some of our percipients have had children as companions, others have had adults; some have had recognized persons, others have had strangers; some have had companions so lifelike that the percipients have never questioned their reality, while others have had what was clearly some kind of paranormal experience.

Have I made the mistake of lumping together a number of phenomena which are really distinct categories? Their many differences might seem to justify us in assigning these companion cases to other categories, were it not that — apart from the 'crisis' experiences of climbers and travellers, which form a special sub-category of their own defined by the unique circumstances — all these companion cases have one factor in common: a relationship that is sustained continuously or intermittently over a period of time, often of many years. We shall find few parallels to this elsewhere in the entity literature.

1.6 ENTITIES AS APPARITIONS

Apparitions are the archetypal paranormal phenomenon. The question 'Do you believe in ghosts?' is the touchstone that distinguishes the sceptical from the credulous. Ghost stories have been told for almost as long as stories of any kind have been told: their mystery has teased man down the centuries, a perpetual challenge that each culture has sought to resolve in its own way, looking for explanations physical or psychological, occult or scientific, religious or anthropological. Each of these approaches has helped us to understand the problem a little more fully, but none has succeeded in furnishing us with a generally acceptable answer. The ghost remains hardly less elusive today than in the time of the ancient Romans.

Complicating the enigma is the fact that apparitions, while sharing sufficient characteristics to seemingly justify their being treated as a single phenomenon, can take a disconcerting variety of forms. Even the most fundamental features of apparitions present us with reasons for not accepting the obvious explanations, because what explains one will not do for another. Consider:

★ Apparitions can be of either the living or the dead. This means that any explanation in terms of returning spirits of the dead can be only a partial explanation at best.

★ Apparitions can be of persons known to the percipient, or they can be of total strangers. This means that any explanation which supposes that they spring entirely from the percipient's own imagination is hard to sustain, without also hypothesizing some process whereby his imagination can perceive the image of someone he has never met in real life.

★ Apparitions can be clearly motivated or seemingly quite arbitrary in their manifestation. This makes it hard to sustain any explanation that supposes a simple cause-and-effect process related to purpose.

★ Apparitions can be vividly lifelike or patently unreal. This presents difficulties when we try to conceive what process is used, forcing us to ask whether the 'unreal' manifestations are merely less successful than the others, or are they trying to do something different, and if so, why?

★ The same apparition (or what seems to be so) can be seen on more than one occasion, and by more than one person, either at the same or another time, which implies some degree of material substance. But they can be seen by some of those persons present yet not by all, which rules out any notion of physical reality as we know it, and implies the possession or lack of an ability to receive on the part of the percipient.

And so on. It seems that whichever clue we take up, it leads us to an impasse where another, contradictory, clue is waiting to prevent any explanation along that particular line.

It would make things easier if we could reject some of the cases and retain only those which did lend themselves to a coherent explanation; but unfortunately there seems no valid ground for dividing the material in such a helpful way. It is true that all apparition cases depend ultimately on the reliability of the percipient, generally with no confirmation from others and almost never with any material evidence; so, depending as they do on personal testimony, it is conceivable that the accounts neglect some vital clue that could provide the key to the problem. But the consistency of such stories over the centuries and the recurrent features reported by so many people in so many different cultures make it difficult to sustain the position that the incidents can be attributed to misperception, errors of observation, misinterpretation of some natural phenomenon, or any other explanation that implies a discrepancy between what is reported and what was perceived. So when a percipient tells us he has just seen the figure of his Aunt Agnes in his bedroom, we have no right to say that he was really seeing a table lamp that he mistook for his aunt, particularly when we later learn that precisely at the time of his sighting his aunt was dying. We have no right to reject his account of the matter unless we have a substantial reason for suspecting him of deception, deliberate or involuntary.

Before we proceed further, let us be quite sure we recognize the implications of the apparition phenomenon by looking at a particular case. I make no apology for offering one of the best known of all such cases, for the reason it is so well known is because it exhibits so forcibly the perplexing features of the phenomenon:

Lieutenant David M'Connel was a British trainee pilot during World War One. He was aged eighteen on 7 December 1918 when he was unexpectedly asked to fly a plane from Scampton to Tadcaster, a distance of ninety kilometres. At 11.30 he left his room-mate, Lieut. Larkin, telling him he expected to be back for tea. Another pilot accompanied M'Connel in another aircraft, in order to bring him back to Scampton after delivering the plane. They ran into fog and landed at Doncaster: they asked for instructions and were told to use their own discretion: they took off again but the companion plane had to make a forced landing. M'Connel reached Tadcaster but crashed on landing and was killed at exactly 3.25 p.m.

At the funeral (11 December) M'Connel's father was told by Lieut. Hillman, another member of the unit, that David's apparition had been seen at the moment of the crash by Larkin. The father wrote to him almost immediately, and on 22 December Larkin replied with an account of his experience.

He had been sitting in front of the fire, reading and smoking, when he heard footsteps coming up the corridor, and then the familiar noise and clatter that M'Connel always made when he came in. Then he heard the greeting 'Hello boy!' Larkin half-turned to look towards the door, which was a little more than two metres behind where he was sitting. He saw M'Connel standing in the doorway, smiling, one hand on the door-knob, dressed in flying kit but wearing a naval cap in place of a flying helmet, as was his habit.

Larkin remarked, 'Hello, back already?' M'Connel replied, 'Yes, got there all right. Had a good trip.' Then with a parting 'Well, cheerio!' he went out and closed the door. Shortly afterwards, at 3.45, Lieut. Garner Smith came in and said he hoped M'Connel would get back early enough for them to go out together that evening. Larkin said M'Connel was already back and had just been in the room: he was sure this had occurred somewhere between 3.15 and 3.30 p.m. Garner Smith confirmed that his, later, visit was at 3.45.

Larkin had no notion that anything out of the ordinary had taken place until he heard about the crash. Then, being sceptical on psychic matters, he tried to persuade himself he never saw M'Connel that afternoon; but apart from his own conviction, there was the testimony of Lieut. Garner Smith to confirm the matter.[148k]

This case — which is merely one of the most straightforward and best authenticated of many thousands of 'crisis apparition' cases — presents the essential features of apparition cases. Though they take many forms, all have this in common: that they appear to be external to the percipient, to be located in space, and to act autonomously. Moreover, they appear to possess identity, even though this may be unknown to the percipient. There is a real sense of separateness, and most witnesses would reject unequivocally the suggestion that the apparition is a figment of their own imaginations for which they alone are responsible. And yet this view is probably that taken by most psychologists.

So long as a purely descriptive account of the sighting is given, there might seem to be some grounds for taking this reductionist view. The ostensibly 'external' location, the autonomous behaviour and supposed identity of the apparition could be no more than illusions, the result of clever tricks: the process involved could be no more complex than that the illusion is created by the right-hand brain, and perceived by the left-hand brain, with no question of involving any outside agency or process. The model is scientifically attractive. Why drag in outside agencies if you can manage without them? Indeed, so far as the *mechanics* of apparition perception are concerned, we shall see that some such model is likely to be valid. And if the mechanics were all, apparitions could be lumped together with hallucinations and most of the phenomena studied in this book as being no more than 'figments of the imagination'.

One definition of the apparition is 'a veridical hallucination', but this is clearly inadequate and unhelpful. For it is evident that there is more to apparitions than the process by which they are perceived. There is also the question of *what* is perceived, and it is in this respect that the reductionist model is so inadequate. The indications of external causation in an instance like the M'Connel case, or at the very least that M'Connel or someone acting on his behalf had a hand in what Larkin thought he saw, are so compelling as to invalidate the 'all in the mind' explanation, at any rate in its simple form. At the same time the implausibility of the incident — why should the just-dead pilot report that he had 'had a good trip' — makes it impossible to accept it as a simple 'visit' by the dead man's spirit.

So let us look at what it is that makes almost all witnesses — and quite a number of theorists — feel that apparitions must have an external cause. Here are four more cases, of widely differing kinds:

★ A German lady wrote to me describing an experience she had while climbing in the Bavarian Alps, when she happened to miss her way. 'You will understand that this is rather a heavy mountain tour, but there is a good way as well up as down, but one must not miss it as I did. Having started a little late for the return, and light beginning to fade, all of a sudden I found myself in a really dangerous position. As a matter of fact one year later a young girl fell to death exactly on the spot where I realized myself to be in an almost hopeless position. All of a sudden I noticed a sort of a big ball of light, and this condensed to the shape of a tall, rather chinese looking gentleman. Extraordinarily I was not a bit frightened, and also not astonished, it all seemed then quite natural to me. The gentleman bowed, spoke a few words, led me a small path to the tourists' way, and disappeared as a ball of light.'

★ In July 1895 Captain Joshua Slocum was in the course of his historic solo voyage, sailing alone around the world. Between the Azores and Gibraltar he ran into squally weather, and at the same time was afflicted with severe stomach cramps which so demoralized him that he went below, not taking in his sails as he should have done, and threw himself upon the cabin floor in great pain: 'How long I lay there I could not tell, for I became delirious. When I came to, as I thought, from my swoon, I realized that the sloop was plunging into a heavy sea, and looking out of the companionway, to my amazement I saw a tall man at the helm. His rigid hand, grasping the spokes of the wheel, held them as in a vice. One may imagine my astonishment. His rig was that of a foreign sailor, and the large red cap he wore was cockbilled over his left ear, and all was set off with shaggy black whiskers. He would have been taken for a pirate in any part of the world. While I gazed upon his threatening aspect I forgot the storm, and wondered if he had come to cut my throat. This he seemed to divine.

"Senor," said he, doffing his cap, "I have come to do you no harm." And a smile, the faintest in the world, but still a smile, played on his face, which seemed not unkind when he spoke. "I am one of Columbus' crew, the pilot of the *Pinta*, come to aid you. Lie quiet, senor captain, and I will guide your ship tonight. You have a calentura (a reference to the stomach cramps) but you will be all right tomorrow . . . You did wrong to mix cheese with plums . . ." I thought what a very devil he was to carry sail. Again, as if he read my mind, he exclaimed, "Yonder is the *Pinta* ahead, we must overtake her. Give her sail; give her sail!" '

Next day Slocum found that the *Spray* was still heading as he had left her — 'Columbus himself could not have held her more exactly on her course. I felt grateful to the old pilot, but I marvelled some that he had not taken in the jib. I was getting much better now, but was very weak . . . I fell asleep. Then who should visit me again but my old friend of the night before, this time, of course, in a dream. 'You did well last night to take my advice," said he, "and if you would, I should like to be with you often on the voyage, for the love of adventure alone." He again doffed his cap, and disappeared as mysteriously as he came. I awoke with the feeling that I had been in the presence of a friend and a seaman of vast experience.'[144]

★ In June 1884 Miss Kathleen Leigh Hunt was staying with her cousin at a house in Hyde Park Place whose owners — relatives of theirs — were abroad.

'One morning after breakfast I was going upstairs when I seemed to see, about two stairs in front of me, a figure, which I took to be the housemaid, going up before me. I went up the entire flight of stairs under this impression to the first floor, when suddenly at the top I could see nobody. This puzzled me as I could not account for anyone being able to disappear so quickly.'

A search revealed that there was nobody about on the upper floor of the house, and she realized that she had seen an apparition. But she insisted that 'the figure itself had nothing supernatural about it'.

After a second, more fleeting glimpse, which again could not have been any occupant of the house, she told her cousin, who said that she too had had similar experiences, and that a former servant has seen 'skirts going up round the doors'.

But, knowing nothing of the previous history of the house, they were unable to identify the apparition. [148b]

★ A Mrs Collyer gave the SPR an account of an apparition which coincided with the death of her son Joseph in a ship collision: 'On the 3rd of January 1856 I did not feel well, and retired to bed early. Some time after, I felt uneasy and sat up in bed. I looked round the room, and, to my utter amazement, saw Joseph standing at the door, looking at me with great earnestness, his head bandaged up, a dirty night-cap on, and a dirty white garment on, something like a surplice. He was much disfigured about the eyes and face. It made me quite uncomfortable the rest of the night. The next morning, Mary came into my room early. I told her that I was sure I was going to have bad news from Joseph. I told all the family at the breakfast table; they replied, "It was only a dream, and all nonsense," but that did not change my opinion. It preyed on my mind, and on the 16th of January I received the news of his death, and singular to say, both William and his wife, who were there, say that he was exactly attired as I saw him.'

Her son Robert, confirming the incident, says, reasonably, 'It will no doubt be said that my mother's imagination was in a morbid state, but this will not account for the fact of the apparition of my brother presenting himself at the exact moment of his death. My mother had never seen him attired as described, and the bandaging of the head did not take place until hours after the accident.'[61]

Each of these very different cases shares the outstanding feature of apparitions — the conviction the percipient feels that he is seeing 'somebody' wholly separate from himself, located at a specific point in space. But to what extent is this conviction warranted? Could the psychologists be right when they suggest that these are mere delusions, formed by the mind, and projected in such a way as to fool the senses?

If so, then the current situation of the percipients would almost certainly be proposed as a causative factor. Two of our cases involve a percipient who is in a difficult or dangerous situation, and the entity appears as a saviour. Did the entities give advice which it was beyond the power of the percipient to give himself/herself? This is debatable. We may feel that if the German lady had known the proper way down, she would have taken it: and even if it was a matter of dredging the information from some deep recess in her mind, there was still no reason to dress the information up in this dramatic way. Why couldn't she just have said to herself, as many strayed travellers have done, 'Oh, it's come back to me now, I should have turned left instead of right at that last fork . . .'? All this business about the materializing oriental gentleman is quite gratuitous on any psychological basis; and if it was her subconscious playing games with her, one can only comment, a fine time to be playing games!

Similarly with Slocum. The topics discussed between the sick sailor and his picturesque visitor were indeed those which he might have been turning over in his mind — the cause of his sickness, the question of whether he had left too much sail on the sloop. Frequent references in other parts of his book show the very great interest he took in Columbus' voyages, so the pilot of the *Pinta* would have been a natural choice if he was summoning up an imaginary figure to keep him company, like the imaginary companions we discussed in a previous chapter. The details — the picturesque clothing, the piratical face — all this would be child's play for his subconscious mind to fabricate from memory and other sources, without supposing anything paranormal.

So Slocum's entity was a likely figure to emerge from his cultural background, to comfort him at such a time; and it may well be that the German climber's oriental friend was similarly rooted in her background. What little I know of her suggests that she may well have been familiar with the concept of the *bodhisattva*, a semi-divine being of the Buddhist belief, who also plays an important part in the doctrine of Theosophy, of which the lady may have been an adherent. But to say that our percipients were conditioned to see these rather than other types of identity does not establish that they were subjective delusions: it could with plausibility be argued that this previous conditioning was what enabled the entities to manifest in their presence!

In any case, no such considerations hold true in our third case. Miss Leigh Hunt was not, so far as we know, in any kind of crisis situation, and in any case the visionary housemaid made no effort to communicate with her, let alone take any positive action or offer advice. If there was any purpose in the incident whatever, it would seem to have been related to the apparent rather than to the percipient. Which implies an external origin, even if we grant that there may have been some degree of 'co-operation' from the percipient — though it is unlikely to have been more than a passive, relaxed state of mind which, as seems to be widely established, is most conducive to this kind of experience.

But it is of course our fourth case, the death-coincidence sighting, which most strongly suggests an external origin. Whatever explanation we come up with has to take into account the fact that, coinciding in time with a crisis in her son's life, a mother came to believe she was visited by his apparition, in a way which involved information to which she could not, by any conventional means, have had access. At the very least such a case — and there are many hundreds of them on record — implies extra-sensory communication: does it imply actual external causation of the manifestation?

An additional complication is introduced by such a case as the following, which occurred in 1926:

Miss Godley, an Irish lady, had been to visit a labourer on her estate who was ill: she was returning home in a donkey trap, her steward leading it and her masseuse walking behind. The road runs along the shores of a big lake, and, while the steward stopped to open a gate there, he asked me, "if I saw the man on the lake". I looked and saw an old man with a long white beard which floated in the wind, crossing to the other side of the lake. He appeared to be moving his arms, as though working a punt, he was standing up and gliding across but I saw no boat. I said, "Where is the boat?" The steward replied, "There is no boat." I said, "What nonsense! There must be a boat, and he is standing up in it", but there was no boat and he was just gliding along on the dark water; the masseuse also saw him. The steward asked me who I thought he was like, I said, "he is exactly like Robert Bowes, the old man". The figure crossed the lake and disappeared in among the reeds and trees at the far side, and we came home. I at once went to take off my hat and coat and to write a note for the doctor, but before I left my room, the bell rang and the doctor came in. I said I was glad to see him as I wanted him to go and see Robert Bowes, he said, "I have just been there," (he went in a car by a different road) "and the old chap is dead." '

Sir Ernest Bennett, who cites the case, includes confirmation from the two other witnesses, though with this discrepancy, that Miss Goldsmith, the masseuse, testified to seeing a boat. [8]

Clearly, the sighting must be linked to the old man's death, which must have taken place at about the same time as they had their entity sighting. Clearly, too, apart from the discrepancy about the boat, all three witnesses were seeing much the same thing in much the same place.

Explanations could be proposed based on various kinds of extra-sensory communication, which would suggest that only one of those present actually had the apparition experience, but that he/she communicated the illusion to the two others. This may be so, of course, though it is only substituting one kind of conjectural notion for another: the only reason for suggesting telepathic suggestion would be to avoid the implications of a multiple sighting. But if we take this case at its face value, then there was a shared crisis-apparition experience, giving us a choice between the following options:

1   Miss Godley, the condition of Bowes preying on her mind, and linking his probably imminent death with the idea of 'crossing the river' to another world which might have been suggested by the lake they are passing, images the scene so forcibly that her two companions pick it up telepathically and are convinced that they, too, see it. The fact that Bowes died at just about the same time was a coincidence, and under the circumstances, not a very improbable one.

2   As above, but Miss Godley is aware by telepathy or clairvoyance of Bowes' death, and it is this knowledge which triggers the hallucination.

3   As above, but it is Bowes himself, perhaps in the act of dying, who projects the image of himself for his recent visitors to 'see'. Miss Godley picks it up and 'transmits' it to the others.

4   Bowes projects a generalized 'death announcement', which Miss Godley picks up and gives a specific dramatized form, which is communicated by ESP as above.

5   Bowes projects an image which is seen by all three simultaneously, so that no ESP between the witnesses is involved.

Since all of these involve some kind of paranormal process, they are all likely to be more or less unacceptable to orthodox science, which would have to find some way of discrediting the testimony itself. But if that testimony stands, then some such explanation as these is called for, though I do not claim to have exhausted our options, and we shall see as this chapter proceeds that there are some models which involve elaborations on the processes just suggested.

At this point we should remind ourselves of a point made earlier, when we were considering hallucinations. It is not the fact of *having the experience* but the *form that experience takes* which takes the entity experience out of the normal run of accepted psychological phenomena and requires us to offer a more sophisticated explanation — not necessarily a paranormal one, though that is really only a matter of definition.

When making this point earlier, I quoted Walter Franklin Prince. Here is the same quotation, but this time continued:

> No species or instance of hallucination can in itself be evidence for the supernormal, however vivid. The evidence, if any, is to be found in the relation of the hallucination to something else, with no discoverable causal link connecting them. The hundreds of recorded instances of apparitions of persons seen at or near the moment of their unexpected deaths at a distance are to the point. In none of these cases is it the hallucination of seeing a person not really present which is reckoned as evidential; the evidence is in the (1) identity of the person (2) with the person then dying or just died (3) without knowledge on the part of the percipient that the person is in any danger of death. Take the case of a man who once in his life experiences an apparition, and it is then of the person whose unexpected death occurs at a distance within the hour — one such case is sufficient to cause an intelligent mind to wonder if there is nothing more than accidental coincidence involved. One such case should not produce any conviction or any reaction stronger than curious speculation. But when such cases have accumulated to hundreds, it should be evident that we have a problem of real importance on our hands, in which mathematics dealing with the calculus of chance must play a part. [123]

The moment we propose an external cause, we presume some degree of reality. We will do well, therefore, to see if we can establish what degree of reality apparitions possess; for though the fact that something *looks* real does not mean it *is* real, the greater the reality, the more elaborate the mechanism required.

The following list of characteristics is adapted from that drawn up by one of the leading theorists in the apparition field, Hornell Hart, who himself acknowledged that his list was adapted from that of predecessors[66]:

1   *Characteristics implying a degree of reality in apparitions*

★ They are generally described by their percipients in much the same terms as other things seen, as if they were real things, and often in considerable detail. Percipients frequently specify that they are as real-seeming as ordinary life, sometimes, even, that they are more vivid.

★ They are often perceived accompanied by other sensory stimuli, they are heard, sometimes they can be touched and felt. And when this occurs, the sensations are generally consistent: if an apparition of a man is seen, then a male voice is heard, and it is heard when he appears to be speaking.

★ There are some cases in which the apparition appears to perform physical actions. Here are two examples:

A subject woke in the night and saw the figure of a strange woman in his bedroom. He spoke, and the figure turned to look at him: at the same time it gave a tug to the roller-blind over the window, which shot up to the roller.[58]

Commenting on this case which they cite in their excellent study of apparitions, Green and McCreery say 'we cannot rule out the possibility that the blind may have been unstable, and released itself in some natural way. Possibly the hallucination was constructed by his subconscious to fit in with the spontaneous release of the blind'. If so, it is very similar to those dream-experiences in which external ingredients — the sensation of cold, the sound of thunder — are built into the scenario.

In another case a Revd and Mrs Gwynne, who customarily slept with a nightlight burning, saw 'a draped figure' passing through their bedroom. The wife 'distinctly saw the hand of the phantom placed over the night-light, which was at once extinguished'. Her husband did not see this action, but agreed that the light, previous alight, was extinguished when the phantom had left. Interestingly, the reason they slept with the light was because they had previously been disturbed by inexplicable sounds in the room, so, unlike the blind case, here it was, in effect, a case of the apparition *causing* the light![61]

★ Apparitions are generally described as resembling persons known to the percipient, or, if not, others who hear the story are often able to provide identification, such as a photograph, which the percipient recognizes. There are often details of clothing to back up the facial resemblance.

★ Reports often include details which the percipient would not be expected to know, but which on investigation turn out to be correct. The case of Mrs Collyer, cited earlier, is a fair example.

★ Apparitions are generally seen in lifelike relation to the visual context in which they are seen; they can be reflected in mirrors or obscured by foreground furniture. Here is a good example, reported in 1893:

Lady B was sleeping in her bedroom with her daughter in another bed at her side. In the middle of the night both ladies suddenly started up wide awake without any apparent cause, and saw a figure in a white garment . . . the room was not quite dark, although there was no artificial light except from the gas lamp in the square outside. The figure was standing in front of the fireplace, over which was a mirror. Lady B saw the figure in quarter profile: her daughter, seeing it from a different direction, saw the back of the figure but not the face, but saw the face clearly reflected in the mirror.[147a]

★ Apparitions can be seen by more than one person at the same time, or by the same person on more than one occasion. In such cases, the sightings appear to be consistent — that is, in the case of a multiple sighting like that of Lady B and her daughter, just quoted, each percipient saw the same figure, in the same place, but at angles appropriate to her viewpoint.

2   *Characteristics implying unreality*

★ Apparitions frequently appear and disappear in a quite unreal way, for example emerging through a wardrobe or a wall, or gradually forming from a mist, as in this case, reported in 1883:

A Miss R had made a compact with a friend, Captain W, that whoever died first would seek to appear to the other. One night, when the Captain was in New Zealand, she woke with the feeling that someone was in the room:
  'I looked about, and presently saw something behind the little table; felt myself grow perfectly cold; was not in the least frightened, rubbed my eyes to be sure I was quite awake, and looked at it steadfastly. Gradually a man's head and shoulders were perfectly formed, but in a sort of misty material, if I may use such a word. The head and features were distinct, but the whole appearance was not substantial and plain; in fact it was like a cloud, formed as a man's head and shoulders. At first I gazed and thought, who is it, some one must be here, but who? Then the formation of the head and forehead made me exclaim to myself "Captain W!" The appearance faded away.'
  She made inquiries about her friend, 'never doubting but that he was dead', but when news eventually came, it was not of his death but of a serious accident in which he had fallen off a coach and been insensible for a while. (Apart from its other interesting aspects, the case raises the question whether, assuming Captain W to have been the agent, he had actually believed himself to be dead!)[61]

The gradual formation of the head may have been deliberate, may have been the only way this particular apparition knew of manifesting: we shall see that a similar process occurs in other types of manifestation, notably spirit seances, and of course it is reminiscent of the lady climber case already quoted.

★ Apparitions are quite frequently seen by one or more of the persons present, yet not by all. In 1900 a case was reported to the SPR in which the figure of a man was seen by different people on several occasions, yet more than once there were people present who failed to see it; here is one instance:

Miss Irvine, a governess, was returning home at about 4.15 in the afternoon, when she was attracted by seeing in front of her a rather tall old man, dressed in a long black cloak . . . She was much interested in this peculiar-looking person, and did not take her eyes off him, whilst she watched him walk backward and forward between the turn of the road and a heap of stones about a hundred yards lower down; he repeated this six times, the last time stopping as if he were speaking to a man who was cutting the hedge at the time. The man who was hedge-cutting did not look round, and seemed quite unconscious of the other's presence. Miss Irvine walked on, and was going to pass the old man, when, to her astonishment, he vanished when she was only about three yards from him.

I know that you will think it foolish of Miss Irvine not questioning the hedger . . . I asked her why she had not, and she answered that she had not liked doing so, as the labourer would undoubtedly have thought her mad, as he clearly did not see any one. [147b]

Several other instances, associated with what was presumably the same ghost, confirmed this 'selectivity', and there are many other cases in the literature.

★ Apparitions do not always respect physical 'laws': they seem to pass through doors and walls.

★ Apparitions occasionally come close enough to the percipient to be touched, but when the effort is made, nothing is usually felt. There are, however, exceptions to this, as we have seen in the case of Emelie Sagée.

★ Apparitions tend to move in an un-lifelike manner, often being reported as gliding rather than walking.

★ Apparitions have been seen moving on a different level than the floor or ground where they are seen. In some cases it was later established that the floor or ground level had been altered at a certain date, and it is supposed that the apparition is of someone who knew the location in its original state, and therefore locates itself accordingly. If this is so, then it is a vitally important clue, for, as we have seen, the great majority of apparitions adapt themselves very well to the circumstances they find, showing themselves aware of their surroundings and of the percipients. That an apparition should not notice that the floor or ground level has been altered implies a quite different *modus operandi*, justifying an explanation along the lines of 'a glimpse into the past'. At the very least, it would seem that two very different phenomena are involved. But before we jump to that conclusion, we need to see the facts established beyond question.

★ Apparitions may communicate with the percipient by paranormal means, employing what seem to be extra-sensory channels. This could be easily dismissed as an illusion on the percipient's part, were it not that on some occasions information is transmitted which seems positive evidence of extra-sensory communication of some kind. The following celebrated case, which occurred in 1876, is hard to explain in any other way:

Mr F G, an American commercial traveller, was away from home on a business trip when one day, at his hotel, at high noon and with the sun shining cheerfully into the room, smoking a cigar and writing out his orders, 'I suddenly became conscious that some one was sitting on my left, with one arm resting on the table. Quick as a flash I turned and distinctly saw the form of my dead sister (who had died of cholera, at the age of eighteen, nine years earlier), and for a brief second or so looked her squarely in the face; and so sure was I that it was she, that I sprang forward in delight, calling her by name, and, as I did so, the apparition instantly vanished . . . I satisfied myself I had not been dreaming and was wide awake. She appeared as if alive. Her eyes looked kindly and perfectly natural into mine. Her skin was so life-like that I could see the glow or moisture on its surface, and, on the whole, there was no change in her appearance, otherwise than when alive.
The visitation so impressed me that I took the next train home, and in the presence of my parents and others I related what had occurred. My father, a man of rare good sense and very practical, was inclined to ridicule me; but he, too, was amazed when later on I told them of a bright red line or scratch on the right-hand side of my sister's face, which I distinctly had seen. When I mentioned this, my mother rose trembling to her feet and nearly fainted away, and as soon as she sufficiently recovered her self-possession, with tears streaming down her face, she exclaimed that I had indeed seen my sister, as no living mortal but herself was aware of that scratch, which she had accidentally made while doing some little act of kindness after my sister's death. She said she well remembered how pained she was to think she should have unintentionally marred the features of her dead daughter, and that unknown to all, how she had carefully obliterated all traces of the slight scratch with the aid of powder, etc. and that she had never mentioned it to a human being from that day to this.[148c]

Apart from the evidence of the scratch, this case seems to support the reality of apparitions: but that of information can hardly have been acquired otherwise than by ESP of some kind though who provided the information, and to whom, offers wide scope for conjecture!.

★ Even when lifelike in appearance, apparitions may behave unnaturally. For example, in the case of Lady B and her daughter, the apparition took no notice of the percipients — yet there it was in their bedroom, having awoken them from sleep! This is so common a feature of entity sightings that it hardly causes surprise: yet for every case such as the 'scratched face' case just quoted, in which the apparition seems to respond to the presence of the percipient, there will be a case in which there seems to be no reciprocated awareness.

These conflicting characteristics make it evident why the problem of apparitions continues to baffle us. Many of them suggest that we are up against a phenomenon which, if not real, yet often makes a considerable effort to be so; yet just as often is content not to be lifelike. Sometimes it gives every appearance of having sought out the percipient and deliberately manifested to him: at others it seems that the sighting is the result of mere chance. That the entity should make any effort at all to appear lifelike is some evidence of purpose on its part: that it should sometimes manifest unrealistically may be a failure on its part, or a limitation to its capability, or simply that some agents are more concerned with verisimilitude than others.

Paradoxically, this inconsistency is a very good argument in favour of an external agency. For if the whole sighting was nothing but pure illusion, there seems no reason why the 'producer' of that illusion should not ensure that the illusion is perfect. To go to the trouble of making the entity look as though it is forming out of mist seems an unnecessary complication.

Again, though the apparition is clearly not subject to the same physical laws that restrict us in this here-and-now world, it shares sufficient of the characteristics of that world to suggest that it is, none the less, a part of it. This is true not only of its appearance, which in most apparition cases seems to be aiming at lifelikeness, but also of the very fact that apparitions manifest at all. Such motivations as they display are similar to those which direct our actions in this mortal life — the communication of information, the utterance of warnings, requests for help and so on; and the fact that these messages are directed at humans implies that the agents responsible for the apparitions are human themselves, or at any rate concerned with human activities. In other words, we are not talking about some free-ranging cosmic force, but about something specifically human-oriented if not itself human.

This is so obvious a point that it is an easy one to overlook: yet it is a necessary first step towards finding out what causes apparitions. In practice, all theories about apparition-origin assume some kind of human connection, and take one of three basic options as a starting point:

1   Apparitions are internally generated, from within the mind of the percipient; this we may term the 'percipient' hypothesis.

2   Apparitions are externally generated, from within the mind of the apparent; this is the 'apparent' hypothesis.

3   Apparitions are externally generated by some third agency distinct from both percipient or apparent; this is the 'agent' hypothesis.

In addition, we may combine two or more of these, hypothesizing some kind of co-operative teamwork.

Consider the most frequently reported type of apparition case. The figure of Aunt Agnes is seen by her nephew George in his bedroom, at a time when they are a thousand miles apart and he has no particular reason to be thinking of her, though, as he subsequently learns, she is in fact at that moment dying as the result of an accident. This is the archetypal crisis-apparition case, of which there are many hundreds on record, so that we may safely discount the possibility that they are all lies or hoaxes or simple coincidences. The idea that chance could be responsible — that George simply 'happens' to call his aunt to mind, and conjures up her lifelike image before him in his bedroom, at a time which just 'happens' to be that of her accidental death, of which he could have had no prior knowledge — is unacceptable except to the most intransigent sceptic. Even though such a coincidence might occur occasionally, it is simply not an acceptable explanation for so many such reports. Nor does it take into account George's own response to the event, which is apt to be that he has had a unique experience, unlike anything he has ever known before, and often of a vividness that is combined with an emotional intensity that seems to come from the entity, rather than to be generated by his own surprise and alarm.

On the percipient theory, Aunt Agnes herself is not involved. So what kind of process can we hypothesize? Could it be that some part of George's mind — and, unless we are to regard him as somehow different from other people, part of all our minds — is continually scanning the universe, extra-sensorily aware of events everywhere? In the process he becomes aware of an event that has a particular significance for him, his aunt's involvement in an accident: his 'monitor' communicates this information to him in the form of a bedroom visitation, conducted in a fairly natural manner.

There is good reason to suppose that part at least of this model is sound. That extra-sensory perception on somebody's part is involved seems confirmed by the fact that what George sees is not always a simple image of the apparent, but a lifelike or impressionistic presentation of the accident itself — he may even see it in considerable detail, including minor details which can be subsequently confirmed as accurate but which he could hardly have known or guessed — for example, that she was wearing a red hat with a feather in it at the time.

We have no evidence that our minds are capable of such a scanning operation, but if such a process could be shown to exist it would account for a wide range of other anomalous phenomena, such as clairvoyance. But there are many objections. For instance, we have to ask why, of all the ways in which George might receive notice of the event, does it come to him in the form of a bedroom visit — an image of his aunt, standing there, often not communicating verbally, and scarcely ever giving the percipient any precise information as to its purpose, and what has happened to the real aunt all those thousands of miles away. Even when the kind of picturesque detail just mentioned is communicated, it is apt to be a stray scrap of information, not like a detail noted in the course of examining a complete picture. All George will get, in the majority of cases, is the figure of his aunt, the red hat, a woebegone look on her face which does at least tell him that something unpleasant may have occurred — and not much more. No indication of time or place, no instructions for him, no real communication such as the real aunt — if she were in a position to do so — would have with her nephew. If we are supposing that his unconscious mind (or whatever else) has managed to get hold of this interesting information, why does it present it to his conscious mind in such an incomplete and unhelpful way? Why is it unable to say to him clearly and unequivocally: 'Your Aunt Agnes has just been killed in a rail accident in the Pyrenees'? Why couldn't Lieutenant M'Connell describe his trip as disastrous rather than 'good'?

Another objection: there must be, at any given time, a great many things happening throughout the universe that could be of considerable interest to George. His Aunt Agnes is in many cases far from being the most important person in his life: why doesn't his 'monitor' bring back news about his girlfriend, or his friend Jim, or his boss, or any other of the people who are of more immediate significance to him? While it is true that some percipients are in a special relationship with the apparent who appears to them, probably a greater number are not, as the cases we have quoted have shown. Often the percipient does not realize or learn till later who the apparent was, if he ever does. There are many cases, for example, in which the percipient is staying in a friend's house and sees the image of a former inhabitant, a complete stranger whose identity he does not learn until later.

Additionally, there is the fact that the critical event — in this case the aunt's death — is more important to the apparent than it is to the percipient. So on all these grounds — that it is *she* who knows about the accident, *she* who knows about the red hat with the feather, and *she* who feels most strongly about the matter — it seems much more likely that it is Aunt Agnes who is responsible for telling nephew George what has happened to her.

Unfortunately, as soon as we start to formulate an explanation on the apparent hypothesis — making the Aunt responsible — we run up against a whole new set of difficulties. And these difficulties are apart from the question of what the actual communicating process is: even if we assume that some process exists whereby Aunt Agnes can transmit visual information in any direction and to any destination she chooses, it is almost impossible to suggest a plausible model.

In the first place, by making Aunt Agnes responsible, we have by no means eliminated the 'cosmic scanning device' with which we had to provide George in order for him to know of his aunt's accident. For if, on the apparent hypothesis, he has no need of it, the same is not true of her, for Aunt Agnes is no more likely to be aware of her nephew's whereabouts than he is of hers. So if she is to manifest in his presence, some part of her mind has to make a lightning scan of the universe to locate him. What is more, she in her turn must have some extra-sensory perception of his situation, for she not only appears to him, but does so in the right physical dimensions, in an appropriate position, standing on the floor between the items of furniture, looking directly at him, as if she were quite familiar with the disposition of his room, though she has never set foot in it, and quite aware of his presence and the fact that he is sitting in an armchair not standing or lying on his bed.

By supposing the apparent to be also the agent, we certainly eliminate the most serious difficulty of why she should appear at all. True, it is not always easy to understand why Aunt Agnes should choose to appear to her nephew George, to whom she was never particularly close, when there are others to whom she was more intimately linked: but one can at least conjecture some explanation for this — perhaps, in such a crisis case, the panicking Aunt Agnes would send out a sort of generalized Mayday call to everyone she knows, and perhaps nephew George was the only one whose mind was in the right 'receptive' mode at the time. Just as a radio set has to be switched on to receive a signal, so we may surmise that the percipient needs to be in a specific mental state to perceive an apparition, even though he may in no way take an active part in causing it to manifest. This supposition is strengthened by 'bystander' cases, in which the person who sees the apparition does not seem to be the person the apparition is intended for. Here is a particularly clear-cut example. A Mrs Clerke reported:

In August 1864, about 3 or 4 o'clock in the afternoon, I was sitting reading in the verandah of our house in Barbados. My black nurse was driving my little girl in her perambulator in the garden. I got up after some time to go into the house, not having noticed anything at all, when this black woman said to me, 'Missis, who was that gentleman that was talking to you just now?' 'There was no one talking to me,' I said. 'Oh yes, dere was, Missis — a very pale gentleman, very tall, and he talked to you, and you was very rude, for you never answered him.' I repeated there was no one, and got rather cross with the woman, and she begged me to write down the day, for she knew she had seen someone. I did, and in a few days I heard of the death of my brother in Tobago. Now, the curious part is this, that *I* did not see him, but she — a stranger to him — did; and she said that he seemed very anxious for me to notice him. [61]

A more difficult question is why at such times the apparent limits the manifestation to a simple appearance. Again we have to ask why, having achieved the remarkable feat of appearing at all, isn't Aunt Agnes more informative or helpful? Consider this case, which occurred in January 1910:

> Mary Travers was sitting up late waiting for her husband George, an insurance salesman, to return home. She heard the clock strike 11; then, some time later, she heard a taxi coming down the street and stopping in front of the house. She heard a voice, presumably the driver's, call out 'Good night!', and heard her husband's familiar steps on the porch. She hurried to let him in: he entered silently, his hat pulled low over his eyes, and stood with his back to Mary while she closed the door. She thought his behaviour strange and asked if he felt all right? He turned to face her — and instead of George's face she saw a strange white death-mask. She screamed, which brought the neighbours hurrying in: the apparition presumably vanished. Minutes later the phone rang to say that George had been killed in a train crash. [42c]

There are many variants of such a story. In a good many the percipient, after identifying the entity, jumps to the quite correct conclusion that 'something has happened' to the apparent, with the result that he is not particularly shocked when the confirming telegram arrives. Such forewarning could be seen as a kindly-meant preparation, but it does not always have that effect. Sometimes the percipient draws the wrong conclusion, as in the compact case quoted above (page 85) where the percipient assumed the apparent to be dead whereas he was only injured. On the whole, nothing seems to have been achieved by the whole exercise, except that a somewhat untrustworthy notification of the event has been conveyed somewhat sooner than it would have been via the postal services.

And if we can raise such objections in crisis cases, where the agent's motivation seems sufficiently strong, what are we to say of those cases where there is no crisis and no apparent motive for the apparition? What of the housemaid seen on the stairs by Miss Leigh Hunt? What of this seemingly meaningless occurrence:

> Mr Rouse, who was closely acquainted with a Mrs W with whom he attended spiritualist meetings, had to go to Norwich one day in 1873, when he would normally have attended one of the spiritualist meetings. Late in the evening he went for a walk in the outskirts of the town: it was a bright moonlit night.
> 'I found myself on top of a small hill, which enabled me to see a considerable distance along the road, the only living object apparently in view being a human form, so far off that I could not tell if it was a woman or a man, and did not take much notice of it. However, in walking on, I soon made it out to be a woman, and concluded it was a country woman walking into Norwich. I began to fear that, the time and place being so lonely, she would be afraid to pass me: I therefore got as near as possible to one side of the

road, but to my astonishment she took the same side as myself as if determined
to meet me face to face. I then walked into the middle of the road, but to my
surprise the woman did the same.

'Instead of a country woman, as I imagined, I could plainly see that the figure
before me was a well-dressed lady in evening dress, without bonnet or shawl.
I could see some ornament or flower in her hair, gold bracelets on her bare
arms, rings on her fingers, and could hear the rustle of her dress. I felt certain
that I had seen the lady before, and immediately afterwards I recognized her
as Mrs W. I had not the least fear, for she was so real that I thought she had,
like myself, unexpectedly and suddenly got to Norwich. I therefore met her
without the least shake or tremble, delighted to see my friend. We approached
within about five feet of each other; she gazed at me very intently as I thought:
she held out her hand to me, and I could see her face and lips move as if about
to speak to me. I was in the act of taking her hand to greet her, but had not
touched her, when some iron hurdles which formed the fencing of the cattle
market rang as if they were being struck with an iron bar. This startled me,
and unconsciously I turned round to see what made the noise. I could see nothing
and instantly turned again to Mrs W, but she was gone.

'Now it was that I began to tremble, and for some time I felt that she was
still near me, though I could not see her. But I soon pulled myself together and
walked back to Norwich and my bedroom, but not to sleep, for I could not
get rid of the feeling that perhaps my friend had suddenly died or met with
some serious accident. I therefore wrote to a gentleman in London — a mutual
friend — telling him what I had seen and asking him to make inquiries. The
next day's post brought me the welcome tidings that Mrs W was quite well and
in good health, that at the very time I saw her, about 11 p.m., she was sitting
in her usual place in the circle in London, and that there, for the first time in
her life, she had fallen into a trance which frightened the other sitters very much,
and they had great difficulty in bringing her back to ordinary life.'

Mrs W confirmed John Rouse's account, including details of what she was
wearing at the time: her subjective impression, on coming round after falling
unconscious, was 'that I had gone suddenly out of myself'.[61]

So there are substantial objections to the *apparent hypothesis*, though
perhaps not as great as to the *percipient hypothesis*. What other options
do we have?

Let us suppose, as many believe, that all knowledge of all events,
past and future, is stored in some vast transcendental data bank, and
that we all, some or all of the time, can obtain access to this store. In
that case, all we need premise is some process whereby Aunt Agnes,
in her distress, sends out some kind of alarm message. This could do
either of two things. It could trigger the emergency services at the data
bank to pass on a message to nephew George, which presents no problem
since the bank not only knows what is happening to Aunt Agnes but
also knows just where George is at this particular moment.
Alternatively, the aunt's distress signal could operate directly on some
receiving device in George's mind, instructing him to make inquiry of
the data bank, whereupon he receives the message as before.

Such a supposition conveniently avoids some of the snags that make
our percipient-based and apparent-based hypotheses hard to accept:

but not only is this great data-bank-in-the-sky utterly conjectural, it in its turn is open to many objections. Who operates it, and why? What form of interface exists between it and our human minds? Why are some incidents and not others communicated, and who is responsible for the selecting — the apparent, the information service, or the percipient? And so on. However, these questions are no more formidable than those that face us in respect of other explanations: and a model on these lines, which has much in common with the Super-ESP hypothesis favoured by many theorists for other anomalous phenomena, must be regarded as a possibility, if only because its logicalness makes it so attractive.

Of all the apparition theories which have been offered to the world, probably none has been found more widely acceptable than that of Tyrrell.[162] Many objections have been levelled at it; but it seems to come closer than any other to the kind of explanation the majority of us would be prepared to accept as viable. It may be briefly summarized as follows.

Each of us has within him a creative potential: Tyrrell terms it 'the producer-level', but I propose to replace this cumbersome term, derived from some schematic model of the mind, with the simpler 'producer', with the proviso that we understand by the word not an external agency but some component of our minds, unknown perhaps to the anatomist but just as biologically integrated as the brain itself.

It is this producer that creates our dreams and other internal visions. And on occasion, Tyrrell suggests, the percipient's producer 'teams up' with the apparent's producer, to put on a joint display which is manifest to the conscious mind of the percipient, a simulated reality, which Tyrrell describes as 'not a physical phenomenon but a sensory hallucination'.

Such a hypothesis satisfactorily accounts for the fact that details known only to the *apparent* are accurately presented, at the same time that the apparition is able to accommodate itself to a 'stage' whose details are known only to the *percipient*. It avoids dragging in any third party, and it covers a wide range of cases from highly detailed dramatic presentations, in which an entire scene is presented with sound-effects and a full supporting cast, to those minimal cases in which nothing more is communicated than a sense of the apparent's presence without any visual sensation whatever. Any shortcomings in the presentation — the incomplete transfer of information, the unreal appearance of the apparent, and so forth — can be accounted for quite simply as defective stagecraft!

However, Tyrrell's model leaves many questions still unanswered. How, in the first place, does the getting-together of the two producers take place? Telepathy, no doubt; but the familiar problems — why *this* incident out of so many, and why nephew George and not somebody else more suitable — persist. Perhaps the analogy of the telephone call is appropriate. It takes two people to make a telephone call. Someone may try to initiate a call, but if there is no reply from the other end, no call takes place. A telephone call about someone's state of health can be initiated either by the sick person phoning the other with a sick-bed communiqué, or the other phoning the invalid to inquire about progress. But once the connection is effected, what started as a one-man effort becomes a two-way affair.

And we can take the analogy further. Just as we occasionally obtain wrong numbers and find ourselves talking with a stranger, so some apparitions miss their mark, with the result that a percipient sees an apparition which is not meant for him at all. And indeed there may be some apparitions which are the result of a panic distress call — the random dialling of someone in the wild hope that someone else, anyone, will respond.

Another helpful approach to the apparition problem was made by John Vyvyan in a fine book which I think may have suffered by its obscure title, *A case against Jones.* [163] Vyvyan suggests that we presuppose two processes that may be involved when percipient B sees an apparition of apparent A. One process is initiated by A, which he calls *psi-projection*. This does not have to be a conscious process, and probably seldom is so; but it is a positive reaching out on behalf of the agent/apparent, which may be presented in naturalistic or symbolic terms. The second process is effected by the percipient, and Vyvyan calls it *psi-perception*. It is simply the action, of which perhaps not everyone is capable, of receiving a psi-projection.

On this assumption, the act of seeing an apparition comprises a psi-projection responded to by a psi-perception, just as many other forms of communication require both a transmitter and a receiver. It seems that some people may be better at psi-perceiving than others, but this may be due to practice, or simply because they more frequently enter the appropriate state of consciousness that enables psi-projections to be received.

The merit of Vyvyan's scheme is that it hypothesizes specific faculties in mankind; the drawback is that the evidence for such faculties is purely circumstantial. Vyvyan would no doubt be the first to admit (and I am the second) that all he has done is to give names to imaginary faculties. But they have the advantage of being neutral names, implying no commitment to any theory beyond the existence of a psi-force which is capable of effecting extra-sensory communication.

In this chapter we have been more concerned with explanations than might seem warranted at this stage of our inquiry, but this has been necessary if we are to establish the complexity of the problem; so long as we do not rush into premature conclusions, no harm should be done. We are now aware of the difficulties attending any explanation that assigns all the responsibility to the mind of the percipient, and equally any that supposes the apparent to be solely responsible. We conclude, therefore, that in some cases at least, some kind of mutual interaction is the most likely explanation for what is taking place.

We also see that we have to presuppose the existence of a faculty or mental component that is capable of receiving information from external sources in some paranormal manner, which it is capable of presenting to the conscious mind of the percipient in dramatic form. It may be that it does all the necessary creative work itself; or it may be that it receives co-operation from elsewhere.

With this enlarged awareness of how complex the problem is, and therefore how sophisticated our explanation of it must be, we can proceed to look at some other kinds of entity-sighting.

1.7 HAUNTING ENTITIES

It is customary to treat ghosts and hauntings together, as two variants of a single phenomenon. But the differences between them are at least as great as those which distinguish apparitions from other types of entity sighting, such as visions and hallucinations. Here is a typical haunting case:

In 1880 a Mrs Brooke, wife of a Captain in the Royal Engineers, went with her five-year-old daughter and a nurse to stay with friends at Wrotham, Kent. She had visited the house before, with her husband, and they had both passed disturbed nights, though they had attributed this to cold rooms and unaired beds. After the first night of her stay, the nurse complained of being disturbed during the night, and said that the servants believed the house to be haunted. This was confirmed by her hostess. That night 'I told the girl to leave the door of her room open and to go to sleep without thinking of any foolishness as of course ghosts did not exist.

'I must insist that I was not in the least nervous or uneasy, only rather curious as to what might happen. I made up the fire, locked the door, and took the further precaution of putting a chair under the handle. At first I thought of sitting up to watch, but, being tired, I at last went to bed and to sleep.

'I awoke after what seemed a short time and heard the clock strike twelve. Although I tried to sleep again I found myself getting colder and colder every moment and sleep was impossible. I could only lie and wait. Soon I heard steps coming along the passage and up the stairs, and as they slowly approached my door I felt more and more alarmed. I scolded myself. I even prayed fervently.

'Then I heard a slight fumbling, as it were, with the handle of the door, which was thrown open quite noiselessly. A pale light, distinct from the firelight, streamed in, and then the figure of a man, clothed in a grey suit trimmed with silver and wearing a cocked hat, walked in and stood by the side of the bed furthest from me, with his face turned away from the window. I lay in mortal terror watching him, but he turned, still with his back to me, went out of the door uttering a horrid little laugh, and walked some paces down the passage, returning again and again. After that I think I fainted . . .'

The following night the nurse came to sleep in the same room, and they both saw the 'grey man', the nurse confirming this in a separate account. The following day Mrs Brooke insisted on leaving, though her friends assured her that the phantom 'only appeared three times and always to strangers, and never did any harm'. It seems that the house had been troubled by these visitations for seventy-five years (that is, since about 1800) and that the ghost was supposed to be of a man who murdered his brother in Mrs Brooke's bedroom and threw his body out of the window. She heard that there was a portrait in existence of one of the brothers, dressed like the entity she saw. [63]

I choose this case because, though it rests on personal testimony only, it is characteristic of a great many such anecdotes in which a stranger has a sighting of an apparition which he then learns has been more or less regularly seen by others in the same location. Though by no means the only kind of haunting case, it is capable of being the most evidential because it involves spontaneous and independent confirmation of a previously reported phenomenon.

Superficially, apparitions and hauntings have one major characteristic in common: both are non-physical sightings whose apparents can often be identified, but who either are dead or can in some other way be shown not to be physically present.

But this similarity conceals an equally crucial difference, which is that ghosts are characterized by the *time* when they are seen and by the percipient to whom they appear, whereas hauntings are characterized by the *place* where they are seen and which they appear to frequent, other factors seeming to be irrelevant or relatively unimportant. Nobody has attempted to lay down a strict definition of how many times an apparition needs to be seen at a certain place to constitute a haunting, but we certainly understand it to have been seen sufficiently often to satisfy us that it is in some way linked with the place where it is seen, rather than with the percipients or the time and date. And this holds true even when only one set of percipients sees it and not others, as when all the members of a family see a haunting figure which is not seen by former or subsequent inhabitants of the house. Generally it is clear enough when a haunting is occurring: the apparition of the same entity, seen over a considerable period in the same place, and by several people, is a clear case of haunting when compared with Aunt Agnes' sudden and one-off manifestation to her nephew George.

When George has that experience, it is evident that the family relationship is a key factor. When it is subsequently established that he saw his aunt at the very time that she was dying, it is no less evident that the time element is also an essential one. But *where* George is at the time of his experience seems to be immaterial: though of course there is no way of saying so for certain, it seems obvious enough that Aunt Agnes might have appeared to him in another place — a hotel room, his office, anywhere; there may be, though, certain limiting factors, such as George's mental state, which make it likely that he would see the phantom in the relaxation of a bedroom rather than the bustle of a public room. Furthermore, such apparitions often manifest in places the apparent never visited in the flesh.

With hauntings it is otherwise. In a case such as the one we have quoted, it would be absurd to pursue any other hypothesis than that the apparition is linked with the place in which it is seen, and further, that this link is the crucial factor. As Mrs Sidgwick, one-time President of the SPR, put it in one of the earliest papers on the subject, any apparition supposes a rapport between mind and mind, but:

> In a haunted house we have a *rapport* complicated by its apparent dependence on locality. It seems necessary to make the improbable assumption, that the spirit is interested in an entirely special way in a particular house, (though possibly this interest may be of a subconscious kind), and that his interest in it puts him in connection with another mind, occupied with it in the way that that of a living person actually there must consciously or unconsciously be; while he does not get into similar communication with the same or with other persons elsewhere. [148a]

Today we should avoid the word 'spirit' and be more diffident about the mind-to-mind connection Mrs Sidgwick assumes, but that is because a hundred years of investigation have taught us not to assume *anything!* Mrs Sidgwick certainly puts her finger on the essential feature of haunting cases, and it is one which will prove the biggest single challenge to a psychological explanation for entity cases. For, whatever conclusions we may come to with regard to simple apparition cases like George and his aunt, there is no question but that in haunting cases the identity of the percipient is secondary, if not totally irrelevant.

True, it may be that only persons constituted in a certain way — 'on the right wave-length', as it is casually expressed — may be able to see a haunting entity: and it is a further possibility that the entity may require some force or energy to enable it to materialize, and this can only be drawn from living persons of a certain constitution. But this is mere speculation, and even if it is so, it would still be the case that it was *what* the percipient was, rather than *who* he was, which is relevant. Consequently we can say, with virtual certainty, that a haunting entity emanates from the apparent and not from the percipient.

It is possible to concoct alternative theories, but only with great difficulty and with little hope of convincing others. For example, we could hypothesize that the first percipient of a haunting entity mistook a natural object — a dressing gown hanging on the door — for a figure which, inspired by emotion, he subsequently concluded must be the spectre of a former inhabitant of the house: and that subsequent percipients are not having apparition sightings at all, but receiving telepathic suggestions from the first and perhaps even from other previous percipients . . . and so on. Few will prefer such explanations, and few will quarrel with the proposition that the most likely explanation for hauntings is that *a person, now dead, has left behind or is projecting his image in such a way as to seem to be revisiting a place with which he was associated during his lifetime.*

The motivation of a haunting is usually ascribed to some strong emotion, such as regret or guilt or sorrow. In stories, if not so often in fact, understanding people from the world of the living carry out some kind of action to satisfy the haunting spirit, or in some other way persuade it to relinquish its activities. This may take the form of carrying out a specified action, such as repaying a debt, which seems to be preying on the poor creature's mind; or talking it into facing up to the fact that it is dead and that it is time to abandon its earthly concerns; or by ordering it to depart in a peremptory fashion, invoking the authority of a god or a demon. Some like to hold spiritualist seances to find out what is going on, and this too seems on occasion to be effective. All this presupposes that the entity is some kind of surviving spirit, which should be moving on to higher things in the next world but is for some reason still 'earthbound'.

This may well not be the case at all. It is by no means established that what is operated on in cases of exorcism and the like is the manifesting spirit. It takes two to make a haunting, the haunter and the haunted; and the effect of, say, exorcism may be to inhibit the *percipient* from playing his part in the proceedings.

There are good reasons for not taking haunting cases at their face value, for they present many puzzling aspects. First, there is the disconcerting fact that haunting entities seldom behave with very much intelligence. Often all they do is to let themselves be seen: they take no calculated steps to help get their purpose across to the percipient and so assist their supposed aims. True, a haunting phantom is often reported as weeping, or seeming to implore: but this is seldom more than a generalized emotion, and it is only rarely that any coherent request or message is communicated. This may of course be because there are obstacles to such communication: but in that case, why does the entity persist in such futile efforts, often over what is to us a considerable time span? (The objection, that time is of no consequence in the next life, is irrelevant, for it is in *our* time-frame that the hauntings occur, as is evidenced by their fondness for the early hours of the morning.) Is there nobody over there to tell the poor thing it is wasting its time? Why are the spirits of the dead given the capability of making this kind of return visit, if they do so to so little effect?

Without knowing the facts, it would be unkind to call ghosts stupid, but it certainly seems as though haunting entities no longer retain the full capacity for reasoning which their living originals possessed (unless, and this is not entirely a frivolous suggestion, the originals of haunting entities were stupid even when alive, and have not stopped being so on being translated to a higher plane of existence). This suggests either that the creature doing the haunting is not the whole persona of the dead person, or that it is not the dead person at all but an impostor. This leads us to speculate whether a haunting entity may be either:

★ a non-human spirit of some kind masquerading as a former human. There is no real evidence that such spirits exist, but if they do, it would explain a great many anomalous happenings, from poltergeist phenomena to phoney communicators at spiritualist seances. We shall be looking more closely at this hypothesis in a later chapter: for the moment, let us just accept it as a possible explanation for haunting.

★ a split-off from the dead individual's personality, rather than the whole of it — a piece that was left behind when the apparent died, and is understandably unhappy about it.

Not for the first time or the last time in this study, this raises the possibility that some of our phenomena may be linked with the concept of divided personality. In this matter, too, we shall be treating the concept in greater detail later, when we have more evidence to take into account, so it would be premature to discuss it too deeply at this stage.

It is worth noting, though, that here we have a possible explanation for one of the puzzles we touched on a moment ago, as to why nobody 'over there' tells the unhappy haunting entity that it is wasting its time. It is widely supposed by those who claim to be able to describe the next life, that the individual who passes over is met by some kind of welcoming committee who explain to him what has happened to him, what is to happen next, and so forth, with the aim of making him feel at home. If something goes wrong — if the dead person has problems in adjusting from one level of existence to another — there are 'emergency services' capable of sorting him out. We may suppose that if the whole individual becomes stuck, the 'rescue squad' swings into action; but if the haunting entity is only a fragment of the individual personality, the rest of which has gone cheerfully into the next world to enjoy or suffer whatever is in store for it, then the 'rescue squad' might not notice that a little bit of him had got left behind.

All of which is so wildly speculative that it would seem like academic game-playing, were it not for the fact that hauntings *do* occur and must somehow be explained. Survival is, if not proved, at least strongly evidenced by some very persuasive findings: and here we do have a plausible explanation for the strange zombie-like behaviour so often reported of haunting entities.

There are other options available to us. For example, it has been suggested that hauntings are merely memory-traces, somehow imprinted on a place by the living persons who once inhabited or functioned there: what is supposed to happen in such a case is that the percipient is, as it were, tuning in to a continuous tape loop, incessantly being performed for the benefit of those who have eyes to see.

There is a similarity here with psychometry, the process whereby certain people, to whom are ascribed psychic abilities, are able to take an object and describe its owner, or provide other information about it or about those who have had dealings with it. There is no doubt that this does occur, and the natural presumption is that the object somehow retains 'memory traces' of its former circumstances, which the sensitive is able to pick up. Exactly what the process is, remains unknown, and it could be that the explanation is along the same lines as one of those we considered in connection with ghosts, that there is some kind of infinite information bank somewhere, and that the object acts as an index reference, enabling the psychic to retrieve the appropriate information from the store.

Whatever the process, the object somehow gives the psychic the necessary lead. And if this is true of grandfather's old watch, then it could equally well be true of the west wing of the old manor house. A case was reported recently in which the walls of a Welsh pub seemed to retain sounds from the past which could be heard under certain conditions: though this does not sound very convincing, there is a plausibility to the idea, and it has often been suggested that haunting entities are somehow 'imprinted' on the place they haunt, an imprint that some people are able to pick up.

As an explanation, it is certainly cumbersome; but then, as we have seen, the phenomenon of haunting is not as simple as it appears at first, and is unlikely to be accounted for in any simple manner. Our study of the mechanisms employed in other kinds of entity manifestation may eventually tip the scales in favour of one or other of these options, or suggest one that we have not yet considered: but the clear indication of an agency external to the percipient, which is given by haunting cases, forces us to recognize a dimension to which hitherto — with dreams, hallucinations and even with apparitions — we have managed to avoid committing ourselves.

1.8 RELIGIOUS VISIONS

This is Catherine Labouré's account of what happened to her on the night of 18-19 July 1830:

The feast of St Vincent was approaching. The day before, our dear Mère Marthe instructed us on the subject of devotion to the saints, and in particular on devotion to the Blessed Virgin. For so long I had wished to see the Blessed Virgin! My desire became even stronger, and I actually went to bed with the idea that I might see my dear Mother that very night. We had each received from Mère Marthe a little scrap from the vestment of St Vincent: I cut mine in half and swallowed one piece before going to sleep, convinced that St Vincent would obtain for me the grace of seeing the Blessed Virgin.

I was sleeping when at eleven thirty I heard my name: 'Sister, Sister, Sister Catherine!' I lifted my curtain and saw a child of about five or six, dressed completely in white, who said 'Come with me to the chapel: the Blessed Virgin is waiting for you!' I was afraid I would be heard going. 'Do not worry,' the child said, 'It is half-past eleven and they are all asleep. Come, I am waiting.' I dressed quickly and followed the child. He went on my left, and from him there came rays of light. To my great astonishment there were lights shining brightly all along our way, and I was even more surprised when the chapel door opened at a slight touch of the child's finger. My amazement was at its height when I saw the candles and torches of the chapel lit as if for a midnight Mass. But I did not see the Blessed Virgin.

The child led me into the sanctuary to the side of the chaplain's chair. I knelt while the child remained standing. Then a moment later, the child said 'Here is the Blessed Virgin, here she is!' I heard the rustling of a silken robe, coming from the side of the sanctuary. The 'Lady' bowed to the tabernacle, then sat down in the chaplain's chair.

I did not know what to do, so the child said again 'It is the Blessed Virgin!' I can't explain why, but it still seemed to me that it was not her that I saw. Then the voice of the child changed to the deeper tones of a man's voice, and he repeated his words a third time. Then I rushed forward and knelt before the Blessed Virgin with my hands on her knees. I cannot express my feelings, but I am certain this was the happiest moment of my life.

The Blessed Virgin spoke to me about my conduct towards my director, and confided some things which I may not reveal. She told me also what to do when I felt distressed; pointing to the altar steps, she told me to come there to refresh my heart, and that there I would find the comfort I needed. I don't know how much time passed. When she left it was like a light going out; she disappeared like a shadow, as she had come. 'She has gone', said the child. Together we went back. When I returned to the dormitory it was two o'clock in the morning.[35]

Catherine Labouré is one of many hundreds of people who have claimed to see entities of a very specific kind. If they are who they purport to be, then this category of sightings may stand apart from all the others in this study because a quite different process would seem to be involved. These entities claim to be, or are identified as being, publicly known figures, usually of religious significance: much the best known, and by far the most frequently seen, is the Virgin Mary, mother of Jesus Christ, who died in the first century AD.

Although in many respects this type of sighting is much like the others under consideration, there are some important distinctive features:

1   The apparent is someone known to the world at large, so that its manifestation has an importance which transcends the individual experience of the percipient.

2   Despite the foregoing, the apparent generally lived so long ago that there is no question of visual recognition, and the identification is based on circumstantial evidence and on inner conviction.

3   The same apparent is seen on many occasions, by percipients living in different parts of the world and at different epochs. Despite many differences of appearance and in the method of manifestation, there is ultimately no doubt of the apparent's identity, though the possibility remains that the identity has been falsely assumed by the entity.

4   In virtually every case, entity sightings of this kind are associated with the Christian religion, and in particular with the Roman Catholic persuasion within that religion. Reports of similar happenings in connection with other faiths, though not totally lacking, are both rare and ill-documented[145] If there is a substantial body of non-Christian vision-seeing, it has escaped the notice of the anthropologists, or they have failed to see it as the equivalent of the Christian phenomenon. Consequently, for practical purposes, we are justified in regarding the Catholic Church as having a virtual monopoly of the phenomenon.

5   The percipient is regarded as having been specially selected, and though the Church insists that this is no matter for pride, it is certainly seen as a cause of honour, and the percipients — at least in the principal cases — have been venerated, even to the extent of being canonized.

6   The sightings are accompanied by manifestations that seem to be paranormal in several instances, ranging from successful prophecies to the revelation of unsuspected springs of water, and are followed by events that again seem paranormal, such as spontaneous cures.

7   There are some characteristics of the sightings themselves which are distinctive:

- an unusually high proportion are multiple sightings, seen simultaneously by two or more percipients.
- a very high proportion of percipients are children, and the majority of the rest are naive or uncultured people.
- the percipients are usually (though not always) devout Catholics, brought up in an unquestioning Catholic environment.
- many of the sightings take place in the presence of a crowd whose members do not share the experience.
- in a high proportion of cases, the percipients have repeated sightings of the same entity over a period spanning several weeks or even months.
- in most cases there is personal communication between the entity and the percipient(s), including messages of a generalized kind ranging from individual admonitions addressed to the percipient, to warnings intended for mankind at large, sometimes including prophecies.

Clearly this spectrum of characteristics amply justifies our placing religious visions in a category of their own. But there is an additional aspect to this type of entity sighting, in that it is apt to occur in a quite different and very specific cultural context. The consequence is that religious visions attract a weight of attention many times greater than that accorded to any of our other categories.

It is easy to understand why. The implication of these alleged visitations is that heavenly beings are taking an active and personal interest in our earthly affairs, and this is of immense comfort to adherents of the Christian faith. As we shall see, there are other belief-systems that make the same promise, but these have yet to attain the degree of credibility Christian belief has secured. For reasons that, though much debated, are yet not fully understood, the mother of Jesus Christ has come to be a focus of intense devotion and expectancy, and these indications that she is responding to these feelings are particularly welcome, as Catherine Labouré's account demonstrates. No other kind of entity-sighting experience carries an emotional charge of such intensity, in which the desire of the individual is supported by a vast communal devotion and by doctrinal approval.

At the same time, and largely as a consequence, religious visions have attracted the attention of commentators and scholars on a scale that one does not find in connection with any other kind of sighting. The literature of ghosts and hauntings, large as it is, is nothing beside the output devoted to religious visions, each of the principal cases generating a flood of studies and tracts. Much of it is of a naivety which is apt to nauseate those who do not subscribe to the faith and dismays many of those who do: even educated and intelligent commentators are liable to abandon their usual standards when writing of Bernadette Soubirous or the Fatima visionaries. One consequence is that commentary which is not partisan on one side is likely to be partisan on the other: the credulity of the believers inspires virulent hostility from the sceptics. Impartial assessment is rarer in this field than in any other: I hope I can manage it in the pages which follow.

Impartiality, in this context, I take to mean, on the one hand, that we must give serious consideration to the claim made by Catholic believers, that some at least of the visionary experiences are what they seem to be, and that it follows that they are quite different in kind from any other entity-sighting experience: in short, that they are miraculous. But at the same time, we must also consider the possibility that the believers are mistaken; that, irrespective of the identity of the entity, no extraordinary process is involved that would justify us in treating religious visions in any way differently from the way we are treating the other phenomena of our study.

The archetype of modern religious visions occurred at Lourdes, in the south of France, between 11 February and 16 July 1858. A thirteen-year-old peasant girl, Bernadette Soubirous, had eighteen separate sightings of an entity she originally described as 'une fille blanche, pas plus grande que moi' ('a girl in white, no taller than myself') but which was subsequently identified — by itself, though indirectly, and by other interpreters — as the Virgin Mary.

The entity spoke with Bernadette, and gave directions for various works to be carried out, including the construction of a chapel in her own honour. It is a measure of the high regard felt by Catholic Christians for this type of manifestation that these requests were fulfilled, and that Lourdes has since become a place of pilgrimage attracting millions of devout visitors. Although others claimed to have had similar experiences at Lourdes at that time, none of these other visions occurred at the same place and at the same time as Bernadette's experiences, even though there were often huge crowds present with her when she had her sightings.

We shall be considering several features of the case in a moment. First, however, let us consider the implications of the fact that the Lourdes experience, and other similar instances, are confined to a particular faith. This in itself prompts the conjecture that this is a cultural phenomenon, limited, for reasons which we shall need to take into account, to a certain category of people in certain circumstances. Evidently there is something about the Catholic faith that provides a favourable climate for such manifestations.

There is of course another way of looking at this. If you hold that the Catholic belief is the one true belief, then you will not be surprised that it is unique in the wonders associated with it. You would find it easy to accept that only those who share the faith would be privileged to see such visions, or receive the benefits of miraculous healing, or be granted messages from the deity. And though there are accounts of such phenomena occurring outside the Catholic faith, they are feeble and rare by comparison.[145] So, by contrast with the other entity phenomena we have studied so far, we must face the possibility that true visions can occur only within the Catholic Church: this is what the percipients themselves, along with the Church to which they belong, believe. For them, these visions are not illusory images to be classed with hypnagogic fantasies or haunting ghosts, but actually are what they purport to be: the figure of the Virgin Mary the percipient claims to see is *really* the Virgin Mary.

Though most non-Catholics, and many Catholics, would discount this possibility, it is in conformity with an item of Catholic doctrine that states that Mary was taken up physically at the conclusion of her earthly life. This doctrine was proclaimed by Pius XII on 1 November 1950: the rationale behind it — insofar as dogma requires a rational basis — is that God would not permit a body, in which he had for a while resided, to decompose. (Even if we grant that there is a God in the first place, and that Catholic theologians have been permitted insight into what God will and will not permit, this is a highly dubious argument: for if all the universe is God's, and no matter ever destroyed, it should be all one to him whether the constituent elements of Mary remain in their earthly form, or as the dust to which the rest of us must return.) As a consequence of the papal dogma, a Catholic is entitled to believe that when a Bernadette sees the figure of the Virgin Mary, she is seeing *the* Virgin, the same one that once, if biblical history is true, walked our earth.

Clearly, then, it is the identity of the apparent that distinguishes visions from other categories of entity experience. If the Catholic theologians are right, and the real Virgin Mary is capable of returning, then this may well be the best explanation for the experience of visionaries like Bernadette. However, we have to bear in mind that not all the apparents who are seen in visions ascended to Heaven in their physical form, as the Virgin is said to have done. Are we, therefore, to distinguish between visions of the Virgin, as seen by Bernadette for example, and those of saints such as Michel, as seen by Jeanne d'Arc? Apart from the theological argument, there seems no reason to do so. So let us pursue the question of identity without relying wholly on the specific theological doctrine that attaches only to Mary.

Proof of identity could consist of direct evidence, such as identification positive and unequivocal, or of circumstantial evidence, such as miraculous events, prophetic messages and the like, which we might feel could only have come from the alleged source. In assessing this evidence we must recognize, as with other categories we are studying, that outward appearances can be deceptive — a fact to which the Catholic Church itself is very much alive: its official attitude to visionary and miraculous phenomena has always been much tougher and reserved than that of the great number of its adherents.

Of all visions of the Virgin, that of Bernadette Soubirous is the one that commands the most widespread assent: that fine Jesuit scholar, Herbert Thurston, rejected many alleged visions and considered many others to be dubious, but held no doubts as to the authenticity of the Lourdes event. Yet even firm believers in the vision are hard put to it to explain why identification of the entity is so equivocal. Bernadette herself, at first, was by no means sure of its identity. She described a girl no taller than herself — and Bernadette was only thirteen years old, and notably underdeveloped as well. The real Virgin may also have been of short stature, of course, but she must have been at least forty-five years old when she ascended to Heaven, and not easily mistaken for a young girl, one would think. One could speculate that in Heaven she was given some kind of rejuvenation process, but such speculation takes us into the realm of absurdity. What does seem certain is that Bernadette's description, as she herself gave it during the early phases of the experience, hardly accords with the general idea of the mother of Jesus.

Nor was the entity very helpful. Bernadette made repeated efforts to learn its identity, but even after her sixth vision she was still calling it 'it', referring to it as *'une petite demoiselle'*, and reporting to a priest 'I do not know that it is she [i.e. the Virgin Mary] . . . she has not told me that'. Only on 18 March, approaching the end of her series of visions, did the entity make any statement as to its identity, and then only in an oblique way: 'I am the Immaculate Conception', a phrase which has been a stumbling-block for commentators ever since. Apart from being nonsense, as phrased, the idea itself alerts our suspicions; for Bernadette's sighting occurred only a few years after the doctrine of the Immaculate Conception had been promulgated by Rome, and it was consequently a notion which was liable to be discussed and preached about. A pious young girl could well have picked up the phrase, consciously or unconsciously, and have uttered it in the illiterate manner she reported of her visionary entity. On the other hand it is hard to believe that the Mother of God, even speaking an unfamiliar tongue, would show such a disregard for grammar.

In other words, the one indication of identity on the Lourdes entity's part is itself a reason for doubting the claim. What of other alleged visions? Taking fifty-seven well-substantiated cases, [160] we find that in thirty of them, the vision does not name itself, while in the remaining twenty-seven it identifies itself as the Virgin Mary in one form or another — 'I am Our Lady of the Rhine' to Joseph Hoffert in Alsace, 1873; 'I am the Virgin of the poor' to Marriette Bées at Banneux, 1933; 'I am the sign of the living God' to Barbara Reuss at Marienfried, 1946, and so on. This is to say that the vision more often than not does not identify itself, and when it does, is liable to do so in some formula which is familiar to the visionary. As evidence of identity, in short, the visions' own statements are worth little.

If the appearance of Bernadette's vision was not what we would expect the Virgin Mary to look like, what of other alleged visions? For the most part they are conventional figures of extremely beautiful young women dressed in fine clothes like the figures of Mary in Catholic churches; but there are often specific differences, particularly in the manner in which they manifest, which ranges from the naturalistic to the stylized — even to the extent of being seen standing on a large white sphere, trampling down a green-and-yellow serpent (Catherine Labouré, Paris, 1830), or in a kind of tableau with illuminated lettering (Pontmain, 1871).[97]

Mary lived so long ago that we have no visual record of her appearance. Catholic theologians claim to know a surprising amount about her history, but even they do not pretend to know what she looked like, what she would be likely to wear or how she would be likely to act. Had all the descriptions given by the visionaries matched sufficiently, and offered some corroborative detail, this would have been supportive evidence. But this is not the case: apart from a general consensus that the apparent is a young and beautiful woman, the descriptions are vague and differ in detail. An impartial verdict has to be that, on the visual evidence, we have no grounds for believing that all the visionaries were being visited by the same entity, or that — in those cases where it identified itself — that it was who it claimed to be.

Apart from the direct evidence, we have various other indications that are often offered as evidence that some at least of the visions are genuine. Of these, the best known are the allegedly miraculous cures occurring at the visionary sites. The question whether such cures have been effected remains controversial,[92, 166] but this is beside the point so far as our inquiry is concerned. Even if it should be established that healings of a nature beyond our present medical knowledge have taken place, this is no proof that the provenance of these cures was the visionary entity, and even less that it in any way validates the identity claims of the entities. The sceptical position has been that no truly miraculous cures have in fact been effected at Lourdes or elsewhere; probably a greater number of people would agree that a pilgrimage to these shrines does in many cases confer physical benefits on the sufferer, but this is by no means the same as saying that a miracle has taken place, and many would attribute these benefits to psychosomatic effects generated by the act of making the pilgrimage.

In several cases, the vision is credited with creating or revealing a spring whose waters are supposed to effect the miracle. This again is a controversial matter. The spring at La Salette, for example, had previously existed, but gave water only intermittently, after heavy rains: after the vision, it is supposed to have flowed on a more regular basis. If this is fact, if the entity indeed caused a physical alteration in the ecology of the place, then this is strong evidence that a miracle occurred, and, miracle-workers being few and far between, represents very strong testimony to the vision's identity. But the facts are not clear: if Mary had truly wanted to convince us of her identity, she surely should have chosen something less ambiguous. In France alone there are at least seventy-five miraculous springs[60]; Mary would have done better if, rather than simply adding to their number, she had chosen something more original, or created a spring in a more useful place, such as the Sahara desert. As it stands, it seems an idea likely to occur to a somewhat more conventional mind.

More convincing from an evidential point of view are the miracles associated with the sightings themselves, and the most impressive of these are related to the fact that many visions of the Virgin are multiple sightings. Three of the best known cases of the present century — Fatima, Beauraing and Garabandal — concern a group of children who had the experience simultaneously, and altogether some 40 per cent of sightings are shared (64 out of 154 listed by one researcher[160]). This is a much higher proportion than for any of the other categories of sighting we are considering in this study, and will be a very significant factor when we come to make our evaluation.

Any multiple sighting raises special problems, for the fact that more than one person sees the apparition suggests that it possesses at least a degree of objective reality. Visions raise the problem in a special degree in that, unlike most other apparition sightings, they often take place — after the first occasion — in the sight of a crowd of observers, many of whom will be on the watch for any sign of signalling between the percipients, or of disagreement between them as to what is seen, or where and when. Under these exacting conditions, the degree of correspondence between the various testimonies obtained at Garabandal, for example, is very impressive: while not totally consistent, the actions and reactions of the percipients indicate a shared experience which makes it difficult to sustain any hypothesis of conscious collusion or fraud.

If we knew that there was such a process as collective hallucination, it would certainly be useful for our purpose: but there is no evidence for such a process, and any hypothesis that assumes it must be conjectural. Hysteria, communicating from one percipient to another, and telepathy, similarly operating from a primary percipient to secondary ones, are two further suggestions which have been made, but in both cases there is insufficient evidence for them to be anything but unproved options. Most of us would feel happier, no doubt, if it could be demonstrated that one of the percipients was the 'leader' and the rest somehow 'followers', but again there is no clear-cut evidence. What we *do* know is that there exists a psychological process termed *folie à deux* which can on rare occasions take the form of what appears, to casual observation, to be simultaneous speech by two or more persons. However, in one of the few such cases to be scientifically studied, reported by Jack Oatman in 1942 to the *American Journal of Psychiatry*, careful monitoring established that one of his subjects was in fact speaking a fraction of a second ahead of the other. While this in no way diminishes the extraordinariness of the phenomenon, it does mean that we can look with some confidence for a psycho-physiological rather than some more esoteric explanation.

Though many visions are shared, it must be significant that they are rarely shared beyond a small closely-associated group. The original group continues to see the vision on successive occasions, but no additional percipients are added to their number, despite the increasingly large crowds of people who gather, crowds which must include people of the utmost piety and devotion, or who can claim psychic powers, or who are intimate with or related to one or other of the percipients. The astonishing happenings at Fatima in 1917, where enormous crowds gathered, is the most striking illustration of this paradox: for, while large numbers of the crowd seem to have had a remarkable visual experience, it was *not* the entity-sighting that was experienced by the original percipients. The solar phenomenon witnessed by so many, though it must have been in some way associated with the entity-sighting (for coincidence is simply too far-fetched an explanation under the circumstances), was by no stretch of the imagination identifiable with what the children were seeing. [21, 101]

Are we to suppose that the percipients were specially gifted? It is unlikely that a group of playmates, in a country village, should all possess and share a gift that is not shared by any of those other thousands — a gift, moreover, which they possessed only during a few years of their childhood. Consequently we must suppose that the ability to see the entity was not something inherent in their individual make-up, but was bestowed on them temporarily. This, of course, is the sort of explanation the Church alleges for its true visions: the Virgin Mary, for reasons best known to herself, selects Bernadette Soubirous, and no other, to be the visionary, and metaphorically opens her eyes, and hers alone, to see her while others do not.

Other cases are more complicated, though, and none more so than the astonishing affair at Zeitoun, a suburb of Cairo, in 1968 and 1969, and sporadically thereafter. Here, the figure of an entity, generally identified as the Virgin, appeared at night, hovering over the roof of a Coptic church; it was seen by a great number of people, though not by everyone who was present, and was even photographed, though the photographic image is never well-defined. One would suppose that so public a case, occurring in our own time, in a major city, and subjected to the degree of media exposure that is now routine, would give the researcher, for the first time, a dossier of incontrovertible facts on which to work: but unfortunately Zeitoun seems to be no less ambiguous than any other case. Once again, if the Virgin Mary was really responsible she continues to keep us all guessing. [97]

Besides the multiple character of the sighting, there is another aspect of visions that distinguishes them from other cases: they are rarely unique events, but generally recur over a period of weeks or months. This is true of most of the outstanding cases, including La Salette,

Lourdes, Fatima, Beauraing and Garabandal. In each of these cases, the same percipients on almost every occasion saw the same entity, for the most part under similar conditions. But on each occasion different things happened or different words were spoken. From the researcher's point of view this recurrence is uniquely valuable, for it means that the percipients' claims could be checked at the time. True, in a sense such checking has proved negative, in that no other person has been able to see what the percipients claimed to see: but this in itself is a valuable clue. The fact that the sightings have been repeated has made it possible for observers, with some pretence to impartiality, to arrange to be present, so that the behaviour of some of the percipients has been recorded in the very greatest detail, with a degree of detachment — something that is true of no other kind of entity-sighting. It is true that even despite this, there are disagreements as to the facts, to say nothing of the way those facts are interpreted. Nevertheless, we do have a fuller record, from which significant items can be usefully extracted.

One of the questions we shall have to resolve before we bring our inquiry to a conclusion is, what *occasions* an entity sighting? In probably a majority of instances, we do not know, and could not know without probing more deeply into the mind of the percipient than circumstances usually permit. What might not have been revealed had Bernadette Soubirous spent a few hours on the psychiatrist's couch!

There is, however, a special category of vision-seeing which offers us a little more information: the visions associated with a process of conversion. Here is a classic instance, the story of Colonel Gardiner's conversion:

This memorable event happened towards the middle of July, 1719. The Major had spent the evening in some gay company, and had an unhappy assignation with a married woman, whom he was to attend exactly at twelve. The company broke up about eleven, and, not judging it convenient to anticipate the time appointed, he went into his chamber to kill the tedious hour with some amusing book or some other way. But it very accidentally happened, that he took up a religious book which his good mother or aunt had, without his knowledge, slipped into his portmanteau . . . He took no serious note of any thing it had in it; and yet, while this book was in his hand, an impression was made upon his mind (perhaps God only knows how,) which drew after it a train of the most important and happy consequences. He thought he saw an unusual blaze of light fall upon the book while he was reading, which he at first imagined might happen by some accident in the candle; but, lifting up his eyes, he apprehended, to his extreme amazement, that there was before him, as it were suspended in the air, a visible representation of the Lord Jesus Christ upon the cross, surrounded on all sides with a glory; and was impressed, as if a voice or something equivalent to a voice had come to him, to this effect, (for he was not confident as to the words) 'Oh, sinner! did I suffer this for thee, and are these thy returns?' Struck with so amazing a phenomenon as this, there remained hardly any life in him; so that he sunk down in the arm-chair in which he sat, and continued, he knew not how long, insensible. [69]

We are not told what happened to the lady friend, or what explanation her would-have-been lover gave her; but from that evening on Gardiner was a changed man. Hibbert, though writing as long ago as 1825, rejects the suggestion that any divine providence was involved in this event; and though he does not say so explicitly, he would surely have agreed with William James [76] that the event was generated by the workings of Gardiner's own subconscious mind. There are abundant case histories to show that guilt and remorse build up over a long period, unknown to the 'sinner', to a point where it only requires the slightest incident to bring about a crisis. In Gardiner's case, the accidental taking up of a pious book was the immediate cause. Hibbert further comments:

> The appearance of our Saviour on the cross, and the awful words repeated, can be considered in no other light than as so many recollected images of the mind, which probably had their origin in the language of some urgent appeal to repentance that the Colonel might have casually read or heard delivered. From what cause, however, such ideas were rendered as vivid as actual impressions, we have no information to be depended on. A short time before the vision Colonel Gardiner had received a severe fall from his horse. — Did the brain receive some slight degree of injury from the accident, so as to predispose him to this spectral illusion?

We have no warrant for supposing that such motivations are at work in every instance of vision-seeing; I think it likely that they may, though not always in so clear-cut a fashion. The little children of Fatima and Garabandal can hardly have been such hardened sinners as adulterous Colonel Gardiner, though the literature of religious experience shows how much guilt can be generated from how small a misdeed.

What I think we *do* have a right to suppose, though, is that the act of seeing a vision is the culmination of a prolonged process, rather than a sudden bolt-from-the-blue event. In which case we may further suppose that, during the gestatory phase, the 'producer' will have gradually gathered together the necessary building materials from which to construct the entity sighting.

Before we attempt to assess the implications of the testimony for visions of the Virgin, we should consider other kinds of visionary experience to see if they throw any light on the phenomenon. These, too, tend to be religious in character; but in other respects they are very different from the apparitions of the Virgin we have been considering.

Let us start, though, with a kind of phenomenon which seems superficially to have something in common with them:

In the mobile home of Mr and Mrs Frank Harley of Pontiac, South Carolina, people line up in the small living room to take turns praying in the bathroom. At night a cross shines through the bathroom window. 'It was a glowing cross, flooding the room with light' . . . Sometimes when you look at it, you can see a burning bush and an angel by the cross.' Mrs Harley had lived in the mobile home for ten years, and the cross had never appeared before. [42 k]

This is far from being an isolated phenomenon; American researcher Scott Rogo has reported on 'the Great Cross Flap' which occurred in 1971, when cases like the foregoing were reported from many parts of America. [135] Nor are they confined to America: a very similar case occurred in Wales a few years earlier. Pauline Coombes, who in 1977 was involved in a UFO case notorious for its ambiguous character, had been living in a trailer-caravan at Pembroke Dock, when she noticed that about 10.30 every evening a figure appeared in the glass of one of the windows. It was a life-size figure, dressed all in white, seemingly in mid-air. Pauline, who was a Catholic, took it to be the Virgin Mary; but in the course of time the figure changed to resemble that of Jesus. The Roman Catholic priest, who came to see it with his entire Sunday School, said, 'Oh, Mrs Coombes, my dear, that's beautiful!'; he was just one of hundreds who came to see it. [124]

Such events hardly constitute entity-sightings of the kind we are concerned with, but they share much of the same enigmatic character. They seem equally arbitrary in their occurrence, and equally ambiguous in their nature. As with visions, they must be considered miraculous if they are what they purport to be: but they lack the force of the visionary sightings — nobody today makes pilgrimages to Pontiac or Pembroke Dock to pay respects to the scene of the strange events. Yet here are things which, supposedly, dozens and even hundreds of people have seen with their own eyes, whereas in the case of, say, Lourdes, the whole matter rests on the unsubstantiated word of a single thirteen-year-old peasant girl.

A special case of vision-seeing concerns sightings that occur at times of national crisis. The best-known example is the epidemic of sightings reported during the early months of World War One, including the celebrated 'Angels of Mons' incident. From many battlefields soldiers reported being encouraged by visionary figures. It was later plausibly claimed by the author Arthur Machen that these tales were based on a fictional story he had written, in which the bowmen of Agincourt appear and inspire the English forces. [99] But as we are already learning in this study, and as cases yet to be cited will confirm, the fact that an alleged entity sighting can be traced to a preceding piece of fiction is not necessarily proof of hoax or of unconscious retelling of the story as fact. In some cases, at least, something more complex is taking place, and this may well be true of the 'Angels'. Machen's explanation, plausible though is sounds, is inadequate to account for the widespread nature of the phenomenon.

Machen's story can only be said to relate to accounts by people who had read, or could have read, his story, or who had somehow been influenced by people who had done so. It hardly accounts for the fact that similar events were taking place in the French and Russian armies: we must suppose that the number of Russian privates likely to have read or heard Machen's story must have been negligible.

World War One got off to a good start for the Germans, and they may well have felt they were about to enjoy the same kind of steam-roller success they had enjoyed in 1870. Possibly the Allies felt the same way for the general view in both France and Britain was that the Germans were certain to take Paris. But then, suddenly and seemingly unaccountably, the German advance came to a halt.

Doubtless there were perfectly sound strategic reasons for this, such as the lines of communication being stretched too far; but this was not apparent to the public at the time. It seemed an astonishing and perhaps, to those in the path of the advancing army, a miraculous thing. Perhaps it was this that fostered the climate in which stories began to circulate that the Allies had had divine aid. It was generally believed (by those on the Allied side) that God backed the Allied cause; such assistance might be miraculous, but it was not very surprising.

On 29 September 1914, a story had appeared in the London *Evening News* entitled *The Bowmen*. The author, Arthur Machen, described how, during the retreat, a company of English bowmen, of Hundred Years War vintage, had appeared in the air, struck terror into the Germans' hearts, and given fresh hope to the Allied troops, who thereupon rallied and turned the tide of the advance. The story was pure fiction, but many people responded to the story with allegedly factual accounts of visionary experiences, involving helpful entities, though these included more religious figures than warriors from history. These reports came not only from the English, but also from the French: Saints George, Michel and Jeanne d'Arc were among those mentioned. What is more, such stories had been circulating in France during August, at the time of the retreat from Mons, *before* Machen's story had been published.

On 28 August, a lance-corporal told a nurse, C.M. Wilson, of seeing:

> a strange light in mid-air . . . I could see quite distinctly three shapes, one in the centre having what looked like outspread wings, the other two were not so large, but were quite plainly distinct from the centre one. They appeared to have a long loose-hanging garment of a golden tint, and they were above the German line facing us. We stood watching them for about three-quarters of an hour. All the men with me saw them, and other men came up from other groups who also told us that they had seen the same thing. [142]

A soldier reported how his regiment had been pursued by German cavalry who had trapped them in a quarry, when suddenly 'the whole top edge of the quarry was lined by angels, who were seen by all the soldiers and the Germans as well. The Germans suddenly stopped, turned round, and galloped away at top speed. One of the Germans, captured, asked his captors who was the officer on the great white horse who led the English, for although he was such a conspicuous figure they had none of them been able to hit him.'

Phyllis Campbell, a nurse, was told by a wounded soldier that he had seen Saint George leading the British at Vitry-le-Francois. Another wounded man confirmed the story, saying 'up comes this funny cloud of light, and when it clears off there's a tall man with yellow hair, in golden armour, on a white horse, holding his sword up . . . before you could say "knife" the Germans had turned.' The English were sure it was Saint George — 'had not they seen him with his sword on every quid they'd ever had?' The French insisted it was Saint Michel, helped in some cases by Jeanne d'Arc, fighting the Germans now and not the English. Phyllis Campbell checked with other nurses and found that all had heard such tales — except the one nurse who had been looking after German casualties.

Confirmation was abundant. A lady, telling the story at a social gathering, added that she found it hard to believe, whereupon one of the officers present said to her, 'Young lady, the thing happened. You need not be incredulous. I saw it myself.'

Similar moral support was being given to the Russians, who on this occasion were our Allies. In October 1914, at 11 p.m. one night, 'a soldier from one of the outposts rushed in and called his captain to witness an amazing apparition in the sky. It was that of the Virgin Mary, with the infant Christ on one hand, the other hand pointing to the west. . . . After a time the apparition faded, and in its place came a great image of the Cross, shining against the dark night sky.' On the following day the Russians advanced westward towards the victorious battle of Augustovo. [142]

A vision of a different sort was reported of the first gas attack, near Ypres in August 1915. The allied troops were inclined to panic as the gas, against which they had no protection, came rolling towards them: but then a figure came walking out of the mist, in the uniform of the Royal Medical Corps. He spoke English with what seemed to be a French accent. He carried a bucket of liquid, and from his belt were suspended a number of small cups which he filled with the liquid and passed round to the soldiers. It was very salty, almost undrinkable; but none of those who drank it suffered any effects from the gas. Afterwards there was no sign of the man. [42h]

Such stories continued throughout the war: as late as Spring 1918 it was claimed that a particularly successful German advance was halted by a troop of 'white cavalry' who could not be harmed by enemy bullets, striking despondency into the Germans. A German officer is said to have reported: 'In front of them rode their leader — a fine figure of a man. By his side was a great sword — not a cavalry sword but similar to that used by the Crusaders; and his hands lay quietly holding the reins of his great white charger as it bore him proudly forward . . . All around me were masses of men, formerly an army, now a rabble broken and afraid . . . We have lost the war, and it is due to the white cavalry.'[42b]

Not surprisingly, the Society for Psychical Research received inquiries about these things, and Mrs Salter, one of its leading members, did what she could to discover what substance, if any, there was to the tales. Her conclusion was that, though no really solid evidence was forthcoming, a number of people really did believe they had had visionary experiences of a supernormal character. This she explained as a kind of hallucination, either subjective or triggered by some external cause, and born out of expectancy, fear or hope.[147d]

(As a postscript, we should perhaps note that in 1930 a German espionage agent, Friedrich Herzenwirth, wrote in his memoirs that the Angels of Mons had been motion pictures projected by German flyers on the clouds, to make the English troops believe that even God was on the German side. They should have known better, that a percipient sees what he chooses to see . . .)

The Angels of Mons, and these related phenomena, form a notable case of individual experiences escalating into a widespread myth. Unfortunately, it is almost impossible to track down the original experiences. It does seem that, whatever part Machen's story may have played in the proliferation of the myth, it was by no means its only source: we may well suppose that there also occurred some visionary experiences comparable to those of the Virgin Mary.

Such 'crisis' visions are not nearly well documented enough for us to make any valid comparisons with the kind of vision we have previously discussed; indeed they are not well enough documented for us to draw *any* useful conclusions. What they do tell us, though, is that if the conditions are right, there can occur a proliferation of sightings purporting to be of a charismatic personage, under conditions which seem both to confer immediate psychological benefits on the percipient, and to carry some kind of message for the world at large. It is almost as though the percipient would, in the normal way of things, have felt presumptuous in allowing himself to see a vision, but at a time of national emergency and military crisis it was somehow excusable.

With the regular Catholic visions, too, we get this combination of personal and global significance, but in a very different form. There is, on the face of it, no question but that the entity has sought out the visionaries, and is using them as intermediaries. It is through them that its warnings — usually generalized threats that unless mankind mends its ways, and pays more respect to God, Mary will not be able to stay her son's hand and dreadful things will take place — are uttered, and also requests made for prayers to be said and chapels to be built.

The messages reported by the visionaries not only do not present us with encouraging evidence of identity, but give us even more reason to doubt. On the one hand there are the exhortations: 'You must recite the Rosary every day in honour of Our Lady of the Rosary to obtain peace for the world and the end of the war for only she can obtain this,' the alleged Virgin said to the Fatima visionaries on 13 July 1917. Then there are the rather distastefully egoistic requests. Three months later the same entity said: 'I am Our Lady of the Rosary: I want a chapel built here in my honour.' It is not easy to believe that the mother of God came down from Heaven simply to say such things, and it would not be difficult to produce grosser examples — the entity which appeared to Adeline Piétoquin at Bouxières-aux-Dames in 1942, for example, gave her advice on politics and the current electoral campaign!

We are fortunate enough to have had a new case occur very recently. It was on 31 October 1981 that fourteen-year-old schoolgirl Blandine Piegay had the first of more than thirty meetings with the Virgin in the garden of her mining village home at La Talaudière. She described her visitor as 'a tall gentle lady, young, very good, wearing a large white and blue veil'. The lady told Blandine, among other things, that she wanted mass to be said in Latin, contrary to recent pronouncements by the Vatican, and that she wanted a basilica built in the Piegays' vegetable garden. She uttered threats — 'If the world continues to offend God, I will no longer be able to stay the hand of my son; there will be punishments' — but she has also given good advice: Blandine gave up eating sweets on the Virgin's instructions.

Without doubt, a great many Catholics refuse to accept that Blandine Piegay was really visited by the Virgin Mary; many would have reservations about the authenticity of all but a few visionary claims. But what criteria can be adduced to separate the genuine from the spurious? Father Herbert Thurston, who spent a lifetime considering the evidence for various mystical phenomena and proved himself a hard-headed debunker of false claims, was dubious about the genuineness of the La Salette vision and downright sceptical of that of Beauraing: yet he unequivocally accepted Lourdes. But at no time did he lay down clearly defined rules for deciding when a vision is what it claims to be. The Church, similarly, went along with public opinion sufficiently to canonize Bernadette, while withholding its approval of the really dubious cases. But what of the 'borderline' cases, which command immense popular support: what of Fatima, concerning which, when an enthusiastic pilgrim cried out 'Long live the Pope of Our Lady of Fatima!', Pope Pius XII said 'That is what I am!'?

The drawback with dogmatic interpretations is that there is no room for manoeuvre; one has to accept whatever alternative dogma prescribes. If what was seen at Fatima was *not* the Virgin Mary, then a Catholic would have to believe that it was a lying spirit, a diabolic impostor of some kind. Even Bernadette at Lourdes had her doubts.

On her second visit to the grotto she took holy water with her, and threw it at the vision, in the belief that if it was an evil spirit, this would effectively drive it away. Even when the entity had passed that test, Bernadette was not completely reassured. On her next visit, she told the vision, 'If you are from God, tell me what you want.'

As is well known, the Church too was by no means inclined to accept the Lourdes vision at its face value. But the escalation of the case was hard to resist, and though the evidence remained flimsy, as we have seen, the Church's resistance weakened. Four years after the event, on 18 January 1862, the Bishop of Tarbes, within whose jurisdiction Lourdes lay, issued the following declaration:

> It is our judgment that the Immaculate Mary, Mother of God, really appeared to Bernadette Soubirous, on February 11, 1858, and subsequent days, to the number of eighteen occasions, in the Grotto of Massabieille, near the town of Lourdes; that this apparition bears all the characteristics of being genuine, and that the faithful are justified in believing it to be true.

But we in turn are justified in asking how hard the Bishop and his advisers worked to establish criteria for the 'characteristics' to which he refers. Accounts of the Lourdes event often omit mention of, or make only passing reference to, the so-called 'false visionaries of Lourdes', of which Jean-Baptiste Estrade, a child at Lourdes at the time, eye-witness of the events then and their historian later, writes:

> A veritable epidemic of visionaries seemed to break out suddenly at Lourdes: it particularly attacked young girls and boys. When some of these children approached the excavations of Massabieille, they fell into a kind of feverish contemplation and saw in the interior of the rocks all manner of phantasmagoric figures. To one fascinated witness would present a Madonna of some kind, decked out in sceptres and crowns; to another a Saint Joseph, with the traditional lily in his hand; this one thought he saw St Peter, that one St Paul, a third the four evangelists. In no time it had become a procession of all the noted saints of Paradise. The beings who figured in these diverse parodies, though bearing a certain artificial beauty, were uneasy, disturbed, and indicated involuntary convulsions which made them repulsive.[36]

Estrade goes on to narrate several examples, of which some were simply silly, others genuinely moving, which seem to indicate that for some of the percipients at any rate some kind of genuine experience had occurred. Estrade affirms that none of them carried conviction as Bernadette had done and, like every other commentator, rejects them all as hoax or hysteria. In this he may well be right: but it is not *self-evident* that this one is false, that one the real thing. The decision is based, at best, on circumstantial factors.

That the mother of a first-century religious leader from the Middle East should manifest herself to a young peasant girl in a distant country some two thousand years later is so very improbable a happening that we are justified in asking for very strong evidence before we accept the proposition. Theologians have proposed grounds for believing this kind of event to be true, and have persuaded many eminent scholars, not to speak of many millions of less informed persons, to share that belief. But in order to do so they have had to fall back on their ultimate argument, which is to assert that there are some events on the religious plane that transcend the supposedly known facts of our physical world, and that this is one of them.

But in such a case, we have the right to require that what *is said to have occurred* should convincingly *be shown to have occurred*. Bernadette's vision should have been, unequivocally, a vision of the Virgin Mary. But, as we have seen, there are very many grounds for doubting that it was. And if that is true of Lourdes, how much more is it true of other cases?

★   ★   ★

There is no question of dismissing religious visions altogether, or even of reducing them to humdrum and banal events. If only from the psychological and sociological standpoints, cases like Lourdes and Fatima are of the greatest interest. Unquestionably, something occurred at Fatima that demands an explanation in terms which outrun conventional science; and there are numerous other circumstances — the communion wafer which was seen to appear on Conchita's tongue at Garabandal, the incorruptibility of Bernadette's body, the accurate prophecies of La Salette — that reinforce our interest.

There is one further dimension to visionary experiences that is of particular interest for a comparative study such as this. The French ufologist Gilbert Cornu has demonstrated that there is a close correlation between visions of the Virgin and sightings of unidentified flying objects. He takes 230 visionary cases that are alleged to have occurred between 1928 and 1975, and plots them on a graph alongside UFO statistics for the same period. A similar piece of research was carried out, on a somewhat different sample, by an Italian UFO journal. [21] Both results show a marked upsurge in visions of the Virgin in 1947, which was also the year in which the current wave of UFO sightings may be said to have commenced. Similarly, there was a marked correspondence with the 1954 UFO wave in France.

The implications are profound; either:

★ the things we see as UFOs are really something associated with the Virgin Mary; or

★ the entity we see as the Virgin Mary is really something to do with UFOs; or

★ there is some common factor that makes some people see, or think they see, the Virgin Mary, while others see, or think they see, UFOs.

If we knew for sure what either of these phenomena was, we should be in a better position to choose between these options. As it is, we must accept the link without understanding it, and use it to help us discover what the phenomena are. But this much is certain: no explanation of either one will be complete unless it accounts for this seemingly bizarre correspondence. In other words, whatever UFO-sightings are, they are something which is related to visions of the Virgin, and vice versa.

This has necessarily been a sketchy outline of the puzzle presented by alleged visions of the Virgin: we have had to ignore many interesting aspects of the phenomena and confine our comment to those most obviously relevant to our inquiry. Speculation and comment on these occurrences already fill enough books to form a library; they clearly represent a significant phenomenon whatever their real nature.

At first sight it is tempting to envy the believer in miracles who is free to think they are what they seem to be. But even the believer will find it hard to believe in them *all*: somewhere he will have to draw the line. But where? How can he say, these were genuine visions of the Virgin, these others are spurious imitations? If Lourdes or Fatima had been unique events, we might have accepted the miracle hypothesis, and put them outside the scope of this inquiry. But they are just two of a huge class, and if they are not *all* miracles, we have to take into account the possibility that *none* of them are. Certainly, nothing we have so far learned about them gives us good grounds for treating them otherwise than as another variety of entity-sighting, presenting some unique features — as each other category has done — but equally bearing on our inquiry.

1.9 DEMONIC ENTITIES

Towards the close of the eighteenth century, a workman of Madrid named Juan Perez was arraigned by the Inquisition for denying that there existed a Devil with power to seize the human soul. This sin he admitted, and explained why:

> After having suffered every kind of misfortune, to my family, my property and my trade, I lost patience and called in despair on the Devil, begging him to avenge me on my enemies in return for my soul and my body. I repeated this day after day, but in vain: no Devil appeared. So I consulted a man who claimed to be a magician, who in turn took me to a woman who he said was more skilled than himself. She told me to go three days running to the hill *des vitillas,* there to call on Lucifer as the Angel of Light, offering my soul, renouncing God and the Christian faith. I did so, but saw and heard nothing. She told me to throw away my rosary and all tokens of Christian belief, to renounce my faith in God, and engage myself in Lucifer's service, acknowledging him as the greater divinity. This I did, but still no sign of Lucifer. The old lady advised me to write a covenant in blood, recognizing Lucifer as lord and master, which I should take to the spot and read aloud. I did it, but it was useless. Then I thought to myself, if there are any devils, and if they are really anxious to get hold of human souls, they will never get a more favourable opportunity than I had given them. Since they hadn't taken me up on my sincere offer, it was evident that no such Devils exist. [55]

Perez's failure to come to terms with the Devil, or even to find one to negotiate with, did not of course prove the Devil did not exist; perhaps he simply did not care to add Perez's soul to his collection. Today we would recognize an alternative option: that practical people like Perez are not the kind to have demonic experiences. Yet humanity's dealings with the Devil often take a very practical form, whether it be the precautions recommended for protecting oneself against the Evil One, or the elaborate ceremonials for calling up and controlling the Names of Power which we shall be looking at in a later section of our inquiry. What could be more practical than the action of Bernadette Soubirous, when she returned with her companions to the Grotto at Lourdes where she had seen a vision claiming to be the Virgin Mary? She took with her a flask of holy water, with which to test the genuineness or otherwise of the apparition. When it appeared, she threw the water in its direction. Later she told her companions that the vision had been not at all offended, but had nodded its head in approval, as though perfectly well understanding that someone in Bernadette's situation would need to carry out such a test.

Apart from what it tells us about Bernadette's vision, the incident is also revealing as to the state of mind of visionaries in general. The fact that a young country girl should have recognized the need for making such a test is significant enough; that she should have known how to set about it reveals how explicitly such ideas were prevalent even among unsophisticated people. To such as Bernadette, their religion was a living, practical reality. To see a vision was one of the things that might happen to you; and when it did, there were steps you took to make sure the vision was what it claimed to be.

Those who do not subscribe to the Christian belief find it hard to understand what it meant to Bernadette or Catherine Labouré to be privileged to see the Virgin; equally, it is difficult for non-believers to accept that for a great many people the Devil — the force of evil incarnate — is a very real personage. Yet there are more stories of his manifestation than of any other entity, the Virgin Mary included.

Of all entities, the Devil is the most protean. Sometimes human in appearance, at other times he is quite monstrous; sometimes handsome, at others hideous; sometimes charming, at others menacing; sometimes arrogant, at others cringing and fawning. It is hard, indeed, even for those who believe in him as a reality, to believe that all these manifestations relate to the same individual: the explanation has to be that the Devil is infinitely devious, and adapts his behaviour, as he alters his appearance, to the particular circumstances. But for so powerful and cunning a personage, it is astonishing how frequently he fails in his ends.

For those who do not believe in him, of course, his many guises and ways of behaving are no problem: indeed, they are part of the solution.

A belief in evil as an actual force is a central tenet in most religious doctrines. No theology has found it easy to get people to accept an omnipotent creator-god who nevertheless permits the unpleasant aspects and injustices of earthly life to occur; but the difficulty is much diminished when an opposing power is introduced, whose aim is to sabotage the good god's efforts by continuous guerilla warfare. Even this does not altogether dispose of the difficulty, for it still has to be explained why an omnipotent god should permit this kind of annoyance: but theologians have been able to construct some very plausible scenarios to justify this, usually involving the notions of free will and redemption. These in turn invite further objections from stubborn sceptics, but by this stage the argument has shifted away from the question of the Evil One's existence, which is certainly a convenient way of accounting for the injustice of human life without having to lay blame on a god you believe to be fundamentally well-disposed towards you.

Aside from such theological subtleties, there are what may be thought of as cultural reasons for believing that demonic powers exist. It is impossible for a hero to display his heroism except by conquering an enemy: he needs a villain, whether it be a tyrannical despot, a dragon, an ogre or a Napoleon of crime. The existence of a devil, whose evil points up the goodness of God, is intellectually satisfying.

On a psychological level, the Devil is useful as providing a scapegoat onto whom we can unload at least some of the responsibility for our wicked doings. Eve could off-load some of the blame for eating the forbidden apple by pointing out that it was the serpent who had tempted her to do so. And so, throughout history, we see wrongdoers asserting that some evil power drove them to it, and as recently as 1981 the 'Yorkshire Ripper' insisted that 'voices' drove him to his crimes.

As a theological, cultural or psychological construct, then, the Devil is certainly a convenient figure, and whether or not the man in the street is convinced by the intellectual arguments for his existence, he is likely to cherish a subconscious feeling that dark forces are responsible for the evil in the world, and that if he is reckless or careless enough, he could get involved with them. The Faust legend is an archetype of wide acceptance.

What concerns us here is whether the Devil is anything *more* than an archetype, a theological construct, a psychological device? Are the demonic entities, which so many witnesses have reported encountering, anything other than mental projections born of private fears?

The early Fathers of the Christian Church seem to have had no doubts of the devil's physical existence: he is continually on hand throughout the wonderful tales of the *Golden Legend*, in which the good never fails to triumph over the bad, but only after vanquishing the sinister schemes of the Evil One who employs every trick in the book to lure or scare the faithful to desert their faith. Hermits are tempted by beautiful seductresses, nuns are terrified by monstrous visitants in their cells. The Devil's powers of impersonation are so great that there often seems no way in which the victim could have recognized his foe, though it has always been held that there are certain tell-tale signs discernible to those forewarned. The prudent Christian was always on his guard and, like Bernadette, would not take even the most plausible entity at its face value.

Bernadette may well have been told stories of the Devil pretending to be a beneficent being of some kind: there are many such instances in monastic history. Fortunately, because it happened so often, the church had provided defensive measures. St Philip Neri, who was widely respected for the ability God had given him to discern spirits in spite of their false appearances, advised spitting on any vision that was under suspicion. One of his disciples, Francesco Maria of Ferrara, had occasion to put this advice into practice. One night he told Philip that he had seen what he took to be the Holy Virgin, all bursting with light. Philip told him that if it came again, he should spit on it. The next night it did indeed come again: Francesco spat, and the entity immediately vanished. However, some time later, while praying, he received another visit: this time it really was the Mother of God who had come to him, but after his previous experience, he was not prepared to take it on trust, and prepared to spit. 'Spit then, if you can,' said the vision. He did his best, but could not raise the least drop of saliva. Whereupon the Blessed Virgin said to him, 'You did well to do as the holy Father advised', and disappeared, leaving her servant inundated with celestial consolations. [129]

We may ask ourselves how grown men, seemingly possessed of their mental faculties, can invent and repeat such tales believing them to be true. Once again, part of the explanation must be the difficulty of drawing a line between genuine and false experiences. If we reject that anecdote, do we reject all such anecdotes? This is more than those who sincerely believe in the physical existence of an incarnate force of evil can bring themselves to do.

Even among educated people, a belief in the reality of demonic forces persists into our own Freud-nurtured day. That bizarre investigator of the paranormal, Montague Summers, held an unquestioning belief in the material existence of the devil. In 1979 a prominent British ufologist, Randall Jones Pugh, announced that he was abandoning his UFO investigations because he had come to believe that by so doing he ran the risk of coming up against actual evil spirits:

> I feel we must accept, for want of a better description, the potential and probable presence of entities (as human beings), whose sole purpose is to destroy straightforward belief in Jesus Christ . . . and that their baleful influence can be traumatic, painful and salutary on those people who accept Christ for what he is and represents. [14]

The French investigator Emile Tizané, a former policeman who has devoted much of his life to the aspects of the paranormal that seem to suggest invading forces — poltergeists, possession and vision-seeing — has come unequivocally to the conclusion that in such cases we are up against evil powers in physical form:

> The living being, the true inductor, is the *secondary cause* of the phenomena whose realization he makes possible, though without understanding them and without having initiated them. Reports of inquiries show that an 'unknown guest' insidiously invades the physiology of a receptive subject, who is in a state of latent disposability, and makes him carry out a series of offences (or even crimes such as arson, to name but one) by means which can be either normal or paranormal, camouflages his actions, sabotages all inquiries, and which vanishes, as it came, when the necessary conditions for manifestation no longer obtain . . . In invading the conscious personality of a living being, this guest awakens in it faculties which it would not otherwise possess. [161]

Coming from so experienced an investigator, this interpretation cannot be lightly dismissed: we must take seriously the possibility that this is what is happening when a percipient reports an encounter with the devil or some other malign force. Tizané, to be sure, is not proposing that the dark forces take on any physical shape — he sees them as invading the victim's mind and operating directly through his subconscious. But if they can do the one, we may reasonably credit them with the ability to do the other, and Tizané might well think this happened in the following case:

> In 1886 a Miss Becket, of the Boston Society for Psychical Research, told how she had tried to hypnotize a friend, Mrs Williams, in the presence of the friend's husband. The two ladies were standing two or three metres apart: in a short while they both became rigid, with arms outstretched. Both turned their heads to watch something which Mr Williams could not see: their faces wore expressions of such unutterable horror that he sprang towards his wife and dragged her to a seat, and needed to use great physical force before he could rouse her from the influence. Miss Becket seemed in part liberated at the same time, but it was a long time before they came fully to themselves. 'When we could talk,' Miss Becket reported, 'we found that we had each seen the same vision, in every detail alike. We both saw suddenly take form out of empty space the giant figure of a man. His face expressed fiendish cruelty and wickedness, and we felt ourselves in part in his power, and knew that he was exulting in this power. He seemed to be followed by a great many pigmy figures, that danced about the room and made ugly faces at us, but dared not do more in the presence of this master spirit. It was when the supernatural malignancy of this frightful creature had almost overpowered us with fear and horror, that our faces expressed such torture as to cause the gentleman to interfere.'
>
> Miss Becket had little doubt what she had seen: 'I have always had a strong faith in religion. My friends were too philosophical to admit dogmas into their minds. The horrible figure must have originated in *my* brain, from its resemblance to *my* idea of a personal devil.' This is confirmed by the fact that Mrs Williams' account is not nearly so sensational. She speaks of 'seeing a strange something — an appearance of a shadowy, transparent film or veil, or sheet of the thinnest vapour, float slowly upward between Miss Becket and myself.'[61]

We may well grant that it was Miss Becket who was responsible for the form of the entity, and the force which caused or enabled it to manifest may equally have come from her; but the fact is that something seems to have been created which had sufficient reality for Mrs Williams to see it also, yet which her husband could not see. Most of us would probably incline to see this case as a graphic example of Eliphas Lévi's noted dictum, 'He who affirms the devil, creates the devil'[94], but it would surely be equally legitimate to assert that 'He who affirms the devil, gives the devil the opportunity to manifest'.

That the forces of evil are able to assume physical shape is certainly accepted by another authority whose practical experience gives him a claim on our attention, the Anglican exorcist Dom Robert Petitpierre.

In practice, the rite of exorcism is not one to which he is often required to resort, for his experience has taught him that there are often simpler and less traumatic ways of sorting out problems in which demonic forces appear to be involved.

Petitpierre suggests[117] that there are four forms of non-human activity that can lead to paranormal manifestations in human life:

1   So-called *poltergeist phenomena,* in which a basically internal situation takes the form of an externalized projection: the result is seemingly paranormal events that can give the appearance of external causation and control — that some kind of spirit is responsible. While he does not think this is generally the case, he would agree with Tizané that when a person is in this state, it could provide the opportunity for an external agency — a nasty spirit, say — to intervene.

2   *Place-memories,* or *imprints,* are 'a natural part of life', Petitpierre suggests. They comprise a kind of relic people leave behind in a place they have inhabited or even simply visited. This is revivable, even after centuries, when conditions are appropriate or if something disturbs the place. They are not harmful or dangerous, but can involve the manifestation of the visible entity of the person with whom the imprint originated. There is, however, no conscious action or purpose: the haunting entity is, to all intents, a zombie.

3   So-called *ghosts* are not easily distinguishable from the imprints, except that in their case there *is* a deliberate intention on the part of the apparent. There is often a specific function to be performed or aim to be achieved. This can be dealt with by sympathetic humans, either by fulfilling the purpose or by persuading the ghost to forget the whole thing.

4   *Demonic forces* proper: these are demons or 'little devils' who are of non-human origin, and whose *raison d'être* is to cause harm or misfortune to humans, twisting us away from good and towards evil. Naturally, Petitpierre sees these in a strictly Christian context: however, even for the non-Christian it should be possible to accept their existence in, say, a humanist context, if he feels so inclined.

Petitpierre's scheme is an interesting and plausible one, and deserves our attention because it is the result not of armchair theorizing but of practical work in attempting to sort out people's very real problems. Each of his categories could be matched up with one of our categories of entities, and would leave open the possibility that the entity was in some fashion external to the percipient and had some kind of physical reality. His demonic entities are not figments of the imagination but actualities of a sort, like that at which Martin Luther flung his ink-bottle or that on which Bernadette sprinkled her holy water.

While it is clearly possible for someone who does not accept the Christian doctrine to believe, none the less, in abstract powers of evil, and, further, in their not-so-abstract manifestation as entities, implicit in this category of beings is a clear-cut conviction that good and evil, as such, exist. Without its evil connotations, the demonic entity is — just another entity. But the difficulty then arises, that good and evil are relative concepts, and often subjectively determined.

In the Portugal of 1917 it was an axiom of faith that what was currently taking place in Russia, the Bolshevik revolution, was unquestionably bad. No doubt many of us would agree, albeit with reservations, and not necessarily for the same reasons; but even so we would find it hard to believe that the Mother of God, even if there should be such a person, should come down to Earth and speak to a semi-literate Portuguese peasant girl and get her to tell the supreme pontiff of her Church to consecrate Russia to the immaculate heart of herself (Mary) with the intention of staving off by this means the unfortunate consequences of the political rearrangement introduced by Lenin and his colleagues.

It is this 'local' quality that effectively discourages us from taking entities like those that appeared at Fatima at their face value; and if this is true of the seemingly 'good' visions then it is no less true of the seemingly 'bad'. When considering the visions of the Virgin Mary at Lourdes, we noted the difficulty of distinguishing between the 'true' and the 'false' visions, and Lourdes, because it is by far the most thoroughly documented of visionary events, gives us a unique opportunity to study the matter. Here is just one of the incidents that occurred there during the time that Bernadette was having her visions:

> A boy named Alexandre, aged eleven or twelve, came back from the grotto and threw himself into his mother's arms. He told her how he had gone with the other children to the grotto, had stopped to pray for a moment, and was standing thinking of nothing in particular, when he saw approaching him a gilded lady all covered with frills. The hands and lower part of the apparition were hidden in a dark cloud like a storm-cloud. 'She fixed me with big black eyes and seemed to want to seize hold of me. I immediately thought she was the ugly one [i.e. the devil] and, without knowing what I was doing, I ran away.' He was trembling all over as he clung to his mother's dress. [36]

Eleven-year-old boys do not normally spend their leisure hours visiting sacred grottoes and praying: the emotional climate at Lourdes at that time was clearly exceptional. In such a climate, extraordinary incidents occurred, generated by the prevailing excitement. Catholic believers and psychologists would agree that a kind of hysterical emotionalism provided the culture for the incidents: where they would disagree would be in the matter of whether any external forces were involved, or whether the incidents resulted spontaneously from the workings of the witnesses' own minds — an issue which arises neither for the first nor the last time in our inquiry.

Few Catholics doubt that Bernadette Soubirous had a genuine vision, that the Virgin truly appeared to her. Few believe that it was she who also appeared to Alexandre or any other of the 'false' visionaries. But if it was not her, who was it? The only being with a motive for impersonating the Virgin would be Satan, whose game, no doubt, was to muddy the waters by throwing doubt on the true vision, thus making it just one among many, of which some were patently spurious. However, in his thoughtful evaluation of the matter, French commentator Léon Cristiani[24], speaking of 'the far too numerous visionaries competing with the humble Bernadette', insists that 'there is always a way of distinguishing the authentic gift, the true charisma, from its demonic counterfeit'. But is this claim justified?

When we read popular accounts of Bernadette's vision, we are given a very simplified picture of what Lourdes was like in the spring and summer of 1858. Fuller accounts reveal that — in Cristiani's words — 'we feel as if we are dealing with an epidemic', and this is borne out by contemporary accounts. 'One has no idea, nowadays, of the credulity that prevailed in Lourdes at that time; people's minds were aflame . . . A crowd of small boys and girls claimed to have seen the Blessed Virgin. A man said to me "My little daughter also sees the Blessed Virgin at the Grotto: so many of them do!" . . . Many of my pupils claimed to have seen visions. They often played truant . . . in their own homes they improvised little chapels . . . I heard a girl of ten or eleven moaning, shouting and screaming in front of the rocky hollow. There, she said, was the Vision. This child was as much honoured as the others . . .' There are many other accounts of individual visionaries who indulged in the most extraordinary antics without discouraging their followers. No wonder that 'even the clearest minds were confused by this plurality of visions, some obviously spurious, and by all these alleged miracles, some of which were more than doubtful'.

Under these circumstances, the government officially requested the Church authorities to issue a statement: implicit in its request was a hope that all the manifestations, the vision of Bernadette along with the others, would be officially condemned by the Church, thus putting an end to all this foolishness. But the Bishop of Tarbes was not to be rushed into any premature ruling, and he nominated a commission to look into the whole matter. As we have already noted, the end result was to approve the vision of Bernadette while disparaging the rest, a conclusion that has been respected ever since.

For the believer, the distinction drawn between Bernadette's vision and the other phenomena is validated by subsequent events. Cristiani writes: 'The reader who has had the opportunity of studying this collection of phenomena may well be astonished at their number, intensity and peculiarity, and may even have formed a rather unfortunate impression, in retrospect, of the appearances to Bernadette. Yet, if he reflects a moment, he will note that still more astonishing is the way all the heavy clouds of doubt and suspicion disappeared quite naturally and simply in the light of truth.' Others, no less reflective, might come to a different conclusion; they might feel that for those who could not bring themselves to accept *all* the phenomena, but were reluctant to accept *none* of them, the official ruling — yes to Bernadette, no to the rest — offered an attractive and convenient solution.

Did the false visionaries see anything? Twenty years after the event, the foremost historian of Lourdes, Father Cros, interviewed some of them. Several were convinced that they had had visions of *something*: one spoke of 'a kind of shadow, but I have no idea whether it was a man or a woman', while another recollected 'a dazzling light, and in the middle of it a rather thick shadow. I could not distinguish any face.' A third recalled: 'I saw a vision, white as a sheet of paper, emerging from the back of the cave — it did seem to have human shape, but I could not make out either face or hands or feet. I saw it come back at least five times. Believe it or not, I did see it, and it was beautiful.' Perhaps the most extraordinary of these remembered experiences is that of a witness who still insisted, twenty years later, that he had seen an apparition, but who now confessed that though he had claimed at the time to have seen the Blessed Virgin, he had in fact seen a man, and, moreover, that the face was not always the same.

Commenting on this, Cristiani says 'Our conclusion could hardly be very different from that reached by Father Cros, who was convinced that Satan was truly the mainspring of all these manifestations, because there was a similarity, a sequence, and one might even say a strategy behind them all . . . Satan cannot fail to sign his own handiwork. By God's grace, there was such a difference between the visionaries, whether women, young girls, or boys and little girls, and the composed and tranquil Bernadette, that no one could be mistaken, and the distinction between good and evil, the true and the false, became self-evident.'

Our conclusion, however, is not necessarily the same. Given the tenets of the Catholic faith, the conclusion is a logical one. But clearly the issue here is one of interpretation. We have already noted, in the previous chapter, that the interpretation of Bernadette's vision is far from being the clear-cut matter it is generally presented as being. In other words, *even if Bernadette had a genuine experience, it was not necessarily a vision of the Virgin Mary.* Consequently, the interpretation of the 'false visions', that they were necessarily of the Devil, is no less in doubt.

We must remember that the 'Virgin Mary' and 'Satan' are theological concepts: any reality underlying those concepts is very largely a matter of faith. For someone who plays so large a part in the affairs of mankind, very little is known about the Virgin Mary, and a great deal of what is said of her is clearly myth and a great deal more can only be labelled theological speculation. But at least we have good reason to believe that there was a historical personage known as Jesus of Nazareth; if so, then he must have had a mother. The evidence for Satan is far less sure.

The literature on the Devil is voluminous: in our own day, a belief that Satan is alive and well and living in California is the theme of a host of books in which lurid case histories, each one a potential basis for an *Exorcist*-style horror movie, are narrated with much relish but with a dismaying lack of precision as regards names and dates. What is especially noteworthy is that virtually all the cases are of *possession;* they seldom include actual sightings, and even when they do, the interpretation is likely to be wholly subjective — the percipient brushes against a stranger in a supermarket, or sees him staring across the street, and somehow 'knows' it is the Evil One. Today the Devil seldom appears in his own guise proclaiming his identity; rather, it seems, he prefers to manifest in the form of a modification of the victim's own personality, leaving it to his advisers to diagnose that there has been a demonic 'take-over'.

A cynic would have the right to suggest that the reason why most contemporary cases of alleged diabolic intrusion are possession cases, rather than apparition cases, is because the former win attention, and with any luck sympathy; the latter are apt to attract scepticism and ridicule. Is this because the Devil you don't see is somehow more believable than the one you do? If so, it sounds rather as though it is we humans who are calling the tune, as it were, telling the Devil we will not believe in him unless he turns up in a believable form, and more or less presenting him with a specification to which we require him to conform.

The matter can be rationalized, of course, without resorting to outright cynicism. After all, if we grant the Devil fairly substantial powers of adaptability and perception, we should expect him to be the first to realize that what was bad enough for a medieval peasant will not suffice for today's victims, who — in Southern California at any rate — are not likely to be taken in by a Devil who is not at least as well-versed in sophisticated mind-control techniques as they themselves are. So today's Tempter is an up-to-date, well-read entity, with the wit to see that appearing in a cloud of smoke and other tricks from his old stock-in-trade will do him more harm than good.

Those who continue to maintain a belief in the existence of a real Devil — and they are surprisingly many, as witness the fringe religious literature of America — have to do so in face of the fact that possession cases are much more readily explainable in other terms than were the entity sightings of former days. When it is the victim himself who is possessed, only dogmatic assertion by the alleged invading entity gives evidence that an invasion has in fact taken place; and since the Devil is the Prince of Lies, why should we take this claim at its face value? It is at least equally plausible to suggest that there was no invasion, nor any invading entity, but that each of us carries about with him his own demon, a kind of secondary personality, representing our evil nature incarnate. A visit from the Devil thus becomes no more than an internal psychological process, albeit modified by external cultural influences.

So — apart from those whose theological belief-system still finds room for a real and objective Evil One — can the rest of us say goodbye to the Devil and all his works so far as visible apparitions and physical manifestations are concerned? I dare say many of us would like to: but there remain some grounds for keeping our options open. There have been few more down-to-earth and practical occultists than Dion Fortune, and her firm belief in the reality of evil spirits must be acknowledged with respect. She identified them as the Qlippoth, the demons of the Qabalah. Qabalist teaching is that each positive Sephira has its negative counterpart, and these evil intelligences are the 'Names of Power' known to magicians. We shall see in a later chapter that some of those who believe in the existence of such beings cannot resist the temptation to seek to harness their mighty forces; but they are also capable of spontaneous manifestation when conditions are favourable — which fortunately, they rarely are. Dion Fortune therefore reassures her readers that in everyday life most of us have little to fear:

> I do not think it in the least likely that the Qlippotic demons will be encountered save through the use of ceremonial magic. They are as rare as anthrax in England, but it is as well to know the manner of their manifestation so that, when encountered, they may be recognized. The great majority of dabblers in occultism are protected by their own ineptitude. They fail to get results, and consequently come to no harm. [45]

If Qlippotic demons are so rare as to be, for all practical purposes, non-existent, does it follow that the general run of seemingly normal people who claim to have suffered a demonic visitation are deluded, or are deluding themselves? Once again it would be tempting to say Yes, and settle for a psychological explanation. But Dion Fortune's Qlippoth are not necessarily the only evil spirits around: perhaps they represent, as it were, the demonic Establishment; but there may also be maverick freelance demons in business on their own account — the elementals, sometimes regarded as no worse than mischievous, but sometimes seen as more dangerous.

Until such time as the existence of these beings can be proved one way or another, there will continue to be disagreement between those who see them as the cause of so many horrific incidents, and those others who prefer a psychological explanation. But the latter will have to acknowledge that the psychological process must be a very complex one. Consider this case, which we might equally well have treated as a haunting case and which, though basically a possession case, became on occasion one of entity-sighting also:

> Magdalene Grombach was a farmer's daughter who worked on the family farm at Orlach, in Wurtemberg. In 1831, when she was nineteen, an outbreak of poltergeist activity commenced, which was to continue for well over a year. In addition, Magdalene saw the apparition of a swathed woman, who gave her warning of an evil spirit who menaced the house and advised that

it should be pulled down. The apparition, who became known to Magdalene as 'the White Spirit', identified herself as a nun who had been put into a convent against her will, and who had sinned in some unspecified but easily imagined manner with a monk who was now haunting the farm; the White Spirit said she would do what she could to mitigate the evil of this Black Spirit, but that Magdalene must be on her guard against it.

Further apparitions manifested of an animal nature, and eventually the monk himself made his appearance. He disparaged the White Spirit and tried to win Magdalene to his side, tempting her in various ways. She always refused to speak to him, responding neither to his promises or his threats. Magdalene was very distressed by all this, often ill and subject to trances: in one of these she reported how the Black and White Spirits had contended together in an unknown tongue. The White Spirit had triumphed in the end, and promised that on the 5th of March following Magdalene's sufferings would end; but until that time she must expect even fiercer harassment from the Black Spirit.

And indeed it came. The Black Spirit persistently appeared at her side, pestering her, but now he went further. Saying 'Now I will enter thy body in spite of thee,' he insinuated himself from the left side of her body. She felt his cold fingers at the back of her neck, and sensed him entering her as she lost consciousness. From then on, during trances which sometimes lasted up to four or five hours, Magdalene displayed all the characteristics of the Black Spirit, speaking in a bass voice, and in terms 'worthy of a demon', in the character of the dead monk. It gave every indication of being a real personality, and for all its evil efforts revealed a great unhappiness, insisting that it wanted to be saved, but doubted that it was possible after so many crimes. But nor could the monk be happy serving the Devil, for 'the worst of it is, that my master has also a master' — above the Devil was God.

It was at this stage that fuller details of the relationship between the two spirits emerged. The nun had been seduced by the monk who had then, to hide his crime, murdered her. The farm building, it was stated, incorporated parts of the monastery where these things had occurred; when they were destroyed, the haunting would cease. Convinced by these circumstances, Magdalene's father had set about pulling down the house. The work was completed on 5 March 1833, when remains of the older building were indeed discovered: on precisely that date, as predicted, Magdalene's afflictions ended and never recommenced. [74]

This brief account does not begin to do justice to a truly astonishing case; but only certain aspects of it are relevant to our present purpose. Evidently, it offers us the components for a number of different scenarios, of which these are I suppose the principal ones, though a number of variations could be added:

1   The 'Black Spirit' entity was an evil spirit, which assumed the identity of a sinful monk to enhance its credibility. According to different theories, the spirit could be either the Devil himself, or one of his minions, or a mischievous elemental.

2   The entity was the haunting ghost of the monk, a real life person

whose wickedness during his lifetime had placed him in the power of the Evil One.

3   The entity was a psychological construct, in which a secondary personality of the percipient assumed the form of the monk, creating from time to time an externalized hallucination of itself when there seemed a need for reinforcement. The 'White Spirit' may well have been another personality, representing opposing forces within Magdalene. The two characters may have been based on actual individuals, but would more likely be artefacts run up by Magdalene's subconscious mind from things she had read or heard about.

To modern tastes, something along the lines of the third of these models is most likely to appeal; but no less than the others, it makes formidable assumptions. To accept the first, we need to believe in the existence of evil spirits. To accept the second, we need to accept not only the existence of haunting ghosts but also that of an Evil Power who forces such ghosts to do his will. Recombine the elements as we may, such assumptions must be made. The third model requires none of these beliefs, but does call on us to accept an extremely complex matrix of psychological processes, possibly involving paranormal cognition into the bargain.

Apart from what we, as students of these things, believe, it is evident that much depends on what the percipient himself believes, or can be led to believe. Many psychologists would argue that the crucial element in a case like Magdalene's is suggestion (including auto-suggestion). This was certainly the view taken by Oesterreich, in his classic study of possession cases, several of which involve not simply modifications of the subject's behaviour, but also claims of entity-sighting experiences. Here is one case he cites:

> A patient had been hospitalized for mysterious fits; the almoner told her it was the Devil who was making her ill. Influenced by this diagnosis, her malady redoubled in intensity and she started to actually see the Devil. 'He was tall, with scales and legs ending in claws: he stretched out his arms as if to seize me; he had red eyes and his body ended in a great tail like a lion's, with hair at the end; he grimaced, laughed, and seemed to say "I shall have her!" ' She was told this happened because she didn't pray enough, so she prayed a lot and even paid for masses to be said on her behalf. However, the more she was talked to about the Devil, the more she saw him and the more frequent and violent her attacks became. Eventually she was transferred to the La Salpêtrière asylum, near Paris. Here, nobody talked to her about the Devil and she stopped going to church so often; as a result she gradually regained her tranquillity, her visions ceased, and eventually she let go of the idea that she belonged to the Devil. [112]

No doubt many of us would like to accept that Oesterreich is right in seeing suggestion as the basic process that results in an entity-sighting: such an explanation authorizes us to reject the concept of autonomously functioning demonic entities altogether. But we must recognize that there are great numbers of people, including many intelligent and educated persons, who take the existence of such evil forces very seriously. Such persons, though they would not necessarily claim that every alleged case of demonic visitation was what it seemed to be, any more than they would claim that every alleged vision of the Virgin Mary was the real thing, would insist that the possibility still exists, not to be ruled out altogether.

As our study proceeds, we shall come across this concept arising in other contexts, and that may help us to decide whether the existence of autonomous evil spirits is something we must accept. But certainly at this stage of our inquiry it would be premature to reject outright a concept that has been widely held throughout the religious history of mankind, and is still fervently adhered to by many millions of people.

1.10 ENTITIES AS 'MEN IN BLACK'

On October 10, 1966, two teenage boys of Elizabeth, New Jersey, reported observing a green-headed being in nearby woods. UFO investigator George Smyth joined a crowd of people who were questioning the lads. As he did so he also noticed a large black car parked a good distance from the crowd. Two dark-visaged, heavy-set men emerged from the car, leaving one of their party seated behind the wheel. The two men joined the crowd and once or twice Smyth heard them question the boys. They had a slight slant to their eyes and spoke with an accent which he was unable to identify. Two weeks after this incident, Smyth received a mysterious phone call. An unidentified voice told him to give up UFO investigation, and then broke the connection. [134]

The reported sightings of 'Men In Black' (MIB) present the entity enigma at its most accessible and yet in some respects its most ambiguous. It is far from clear that the entities exist at all, yet the reports we have of them describe them in terms far more naturalistic than those used by percipients of any other category.

MIB are not described as mystical or superhuman visitors from another dimension, though they are often credited with non-human attributes. They are apt to pass themselves off as UFO investigators acting on behalf of government or semi-official institutions such as the CIA in America or the Ministry of Defence in Britain. Their motives are clearly expressed — usually either to extract information from the percipient or to warn him to hold his tongue. Their contact with the percipient is on a much more coherent, personal level than that of almost any kind of entity. At the same time, their personification nearly always contains some anomalous feature that makes the percipient suspicious of their identity.

It is evident that the MIB have already acquired some degree of mythic status. It may indeed be that they exist on no other level, that they are creatures of fantasy, projections of fear or distrust. But if so, it would seem to be a commonly held fear, for the MIB have a strong corporate identity: if they are hallucinations, they are shared hallucinations, being seen by different people on different occasions in very similar form. Consequently, even if for no other reason, the MIB must be of the greatest interest to the sociologist and the mythologist as an example of contemporary folklore in the making. For this reason I propose to give them somewhat greater attention than they intrinsically merit; by giving us the opportunity to observe a stereotype in the actual process of formation, the MIB may throw light on other entities — for example, visions of the Virgin Mary — in which a number of recurrent characteristics seem to have acquired the status of criteria.

If the MIB were no more than contemporary folklore, we would have to regard them as somewhat marginal to the main thrust of our inquiry: but we are in no position to limit them to that extent. However different a masquerading UFO investigator may seem to be from a transcendent vision of the Mother of God, the sightings have sufficient in common for us to recognize them as comparable phenomena:

★ Both types of entity are seemingly real to the percipients at the time of sighting, but are recognized, then or subsequently, as displaying 'non-real' features in their appearance or behaviour.

★ Both tend to appear to the percipient when he is solitary, choosing their time with a care that suggests intelligent planning.

★ Both entities refuse to leave behind any tangible proof of identity or existence. Verbal self-identification is frequently ambiguous.

★ Both are apt to behave in ways that defy natural laws — for example, appearing and disappearing in an 'impossible' manner.

★ Both have sufficient plausibility — within the percipient's frame of reference — to lull his scepticism, at the time at any rate. Thus, religious visions tend to appear to believing Catholics, MIB to UFO witnesses or investigators who take the UFO phenomenon seriously.

★ Both display ambiguous features that make it clear to objective assessment (which may be that of the percipient himself at a later time) that the entity is almost certainly not what it purported to be at the time. We have noted how visions of the Virgin tend to indulge in unseemly political partisanship; similarly the MIB adopt unconvincing attitudes that negate any factual elements in their 'story'. One example of this is their proneness to imply that, as officials, they have authority to enforce their requests; but there is no case on record in which their threats have in fact been made good.

As we shall see, there are persuasive grounds for believing that the MIB, like many of the other entities we are studying, are archetypal entities wearing a new set of clothes appropriate to the context in which they appear. Attempts have been made to show that there have been

MIB throughout history, and that the latest incarnation differs from the predecessors only in superficials. While this may be so, it begs the question of what we mean by an 'archetypal entity'. Do we mean a 'real' creature that has adopted a particular appearance the better to convince us of its existence, or do we mean some kind of projection from within our own minds? The MIB present this dilemma with a fascinating literalness.

The typical MIB report runs as follows. Some time, usually quite soon after he has seen (and most usually reported) a UFO, the percipient receives a visit. Often it occurs so soon after the sighting that there has been no time for any report of the event to have been published: that is to say, there is no explicable way in which the visitor(s) could have obtained access to the information.

Almost always the percipient is alone when the visit occurs, and the timing of the visit is often so well chosen as to make chance unlikely. (In the Hopkins case, which we shall consider in detail in a moment, the MIB arrived when the percipient's family had gone off to a movie and left him alone at home.) The visitors, usually three in number, arrive in a large black car: in the United States this is most often the prestigious Cadillac; in Britain a Rolls Royce or a Jaguar. Seldom, though, is it the latest model: but, while not being up-to-date, it is immaculate in condition and may even look and smell new. If the percipient notes the registration number and subsequently checks it, the number is invariably found to be non-existent or unlisted.

In the majority of cases the visitors are male: only very rarely do they include a woman, and never more than one. In appearance they conform closely to the stereotype of a secret service man. They wear dark suits, dark hats, dark shoes and socks, but white shirts: percipients frequently remark on the immaculate state of their clothing — suits sharply creased, linen spotless.

The physical appearance of the visitors is frequently described as vaguely 'foreign', and most often as 'oriental' — slanting eyes are a typically remarked feature, and dark-hued or heavily tanned skin is *de rigueur*. Sometimes there are more bizarre features: the Hopkins visitor seemed to be wearing lipstick. The MIB are generally unsmiling and expressionless, their physical movements stiff and awkward. Their general demeanour is formal, cold, sinister, even menacing: no warmth or friendliness are shown, though signs of outright hostility are also seldom encountered. Witnesses often express, looking back on their encounter, a sense that their visitors were not human at all.

Some MIB proffer evidence of identity, and may even wear uniforms. They may produce identity cards; if they do give names or other evidence of their status, these are invariably found later to be false: if the MIB identifies himself as an Army officer, the Army will deny having such an officer. This, of course, is no proof that the claim is false: it may be Army policy to deny such claims even if true.

The motivation is clearly expressed, and is usually either to obtain information or to issue a warning. When asking questions, the visitors are well informed; they clearly know of the percipient's experience, and often know enough about him to suggest that they have been checking up on him. If the percipient has been in touch with investigators, they will know about that, too.

The MIB tend to speak in stiff, formal phrases; this is particularly noticeable when they issue threats to the percipient about the consequences of withholding co-operation. Such lines of dialogue as: 'Again, Mr Stiff, I fear you are not being honest', or 'Mr Veich, it would be unwise of you to mail that report' (these are both genuine examples), suggest that the MIB acquire their knowledge of English usage from studying Hollywood B-movies.

The visit almost invariably ends with a warning to the percipient not to tell anyone about the visit, if the percipient has himself sighted a UFO, or to abandon the investigation, if he is an investigator. Violence is often threatened, so the MIB are sometimes referred to as 'the silencers'.

The scenario just outlined is, of course, never played out in its complete form. An analysis of thirty-two of the more detailed and reliable cases shows that they all diverge substantially from the stereotype. For instance, in four of the thirty-two the visits were not made in person, but were limited to telephone calls: in only five cases were there the full complement of three visitors, four in two cases, two in five, and in the remainder just one.

Similarly with the picturesque details. Though in the United States, where the majority of MIB cases have been reported, the car is the commonest means of transportation, in only nine of the thirty-two cases are cars specified (this does not of course mean that they were not used in the others, simply that they are not mentioned in the report). Of these nine, only three were the stereotype Cadillac: only two were specified as black and only two — not the same two — as out-of-date models. In other words, here we have a fine example of the myth-making process at work: the 'best' details tend to be accepted as the norm, even though they do not predominate numerically.

The most publicized MIB case is that of Albert Bender, director of the International Flying Saucer Bureau. Despite its grandiose title, this was just one of many American UFO-investigation groups that proliferated during the 1950s, and its headquarters was Bender's own home at Bridgeport, Connecticut. Formed in 1952, at the height of the 'flying saucer' cult, it quickly attracted a fair-sized membership: and then, after only about a year of operation, it was disbanded, allegedly due to pressure from three sinister visitors who came to Bender's home.

Bender claimed that he had discovered the 'secret' of the saucers, and announced that he was about to publish it in IFSB's journal *Space*

Review. Before finally committing himself, he felt he should test out his ideas on a colleague: he mailed his report off — and a few days later the MIB came calling. In his account[7] he tells how he was lying down in his bedroom, after being overtaken by a fit of dizziness; suddenly he became aware that 'three shadowy figures' had entered his room.

> The figures became clearer. All of them were dressed in black clothes, they looked like clergymen, but wore hats similar to Homburg style. The faces were not discernible, for the hats partly hid and shaded them. Feelings of fear left me . . . The eyes of all three figures suddenly lit up like flashlight bulbs, and all these were focussed upon me. They seemed to burn into my very soul as the pains above my eyes became almost unbearable. It was then I sensed that they were conveying a message to me by telepathy.

The message they thus communicated was that, yes, Bender *had* stumbled upon the correct solution to the UFO mystery: but they forbade him to reveal it. Further, he was to disband his organization and cease publication of its journal. Because he had managed to learn so much of the truth, they were kind enough to tell him the rest: this so frightened him that he was only too happy to conform with their demands. He was instructed not to tell the truth to anyone, and this he agreed to do 'on his honour as an American citizen'.

What immediately arouses our suspicions is that, despite this experience, Bender nevertheless felt called upon to write a book-length account of his experiences. If he really had had so overwhelming an experience, to explain his conduct would, one might think, have seemed absurdly trivial: Bernadette Soubirous, for example, did not feel called upon to write a popular account of her adventure (though what a best-seller it would have been!). Nor does Bender's 'tone of voice' make his account more convincing, though we should not be surprised to find his experiences described in the language of contemporary American popular journalism: its clichés and colloquialisms were, after all, as natural to him as more elegant phrasing would have been to, say, a witness reporting a ghost to the Society for Psychical Research in the 1890s.

And indeed, the entire Bender affair must be judged in the context of a particular epoch, the 'age of the flying saucer'. Almost from the start, the UFO attracted a form of attention that tended to obscure its very real scientific importance. A lunatic fringe formed at the periphery of the subject, subscribing to the belief that UFOs are alien spaceships crewed by denizens of other worlds who are visiting our Earth out of concern for the sorry state of our civilization. We shall be looking more closely at this aspect of the matter when we consider UFO-related entities *per se*; but for now it is important to relate the MIB to the same climate of opinion.

Is it just an accident that the MIB are so closely related to ufology? Almost certainly not: but the relation may be less one of cause-and-effect than that both reflect a common state of mind. A great deal of ufology is conducted in a paranoid mental condition (which of course may be justified!), and the MIB can be seen as expressions of this paranoia. As such they can be interpreted as bogeymen playing much the same role as the devil in theology, or as kinsmen of the innumerable mythological entities who recur in various guises throughout the world's folklore.

However, while we need not doubt that MIB play a comparable role to their predecessors, that is not quite the same as saying that they are necessarily 'the same thing' wearing new clothes. There are very substantial differences — principally, the marked degree of natural-seeming of the MIB encounter. The majority of MIB reports make it clear that the percipient, at the time, took his visitors to be normal flesh-and-blood humans, and only later came to think otherwise. Here is a typical case:

On 13 July 1967 Robert Richardson, of Toledo, Ohio, was driving at night when, coming round a corner, he was confronted with a strange object blocking the road. He was unable to stop in time and hit the object, though not hard. Immediately on impact the object vanished into the air. Police who inspected the scene could find only Richardson's skid marks as evidence of the incident, but when he himself visited the site a second time, he found a small lump of metal which he thought might have been knocked off the object during the collision. He did not tell the police of his find, but sent the material to APRO, one of America's leading UFO investigation groups.

Three days later, at 11 p.m., two men aged in their twenties called at Richardson's door and questioned him for about ten minutes about the incident. They did not identify themselves and Richardson — to his own later surprise — did not ask them to do so. They were not unfriendly, or unduly sinister: they uttered no warnings or threats, but confined themselves to questions. He noted that they left in a black 1953 Cadillac — that is, a model which was fourteen years old: the number, which he noted and subsequently checked, was found not yet to have been issued. This alone indicates that whatever his visitors were, they were certainly impostors of some kind; but at the time Richardson had no feeling that they were anything but human.

A week later he received a second visit, from two different men who arrived in a current model Dodge. They wore black suits and had dark complexions: although one spoke perfect English, the second had an accent, and Richardson felt there was something somehow foreign about them. At first they seemed to be trying to make him admit that he had not had any kind of unusual experience, but then they asked him to hand over the piece of metal that he had found. When he told them that he had passed it to APRO for analysis, they threatened him: 'If you want your wife to stay as pretty as she is, then you'd better get the metal back!' — again an echo of Hollywood script-writing!

At this time, only four people could have known of the existence of the metal fragment: Richardson himself, his wife, and two senior members of APRO. Seemingly, the only way in which the strangers could have learnt of its existence would have been by tapping either his or APRO's telephone. There was of course good reason to suppose that the CIA or some such body might have been listening in to APRO's phone calls, and if so this would let us out of having to impose a psychic interpretation. There was no indication that the two sets of visitors were connected with each other, or even knew of the other's existence: but both clearly had access to information that was not publicly available. [95]

The hypothesis that the MIB are CIA agents or something similar is perhaps the most obvious one, and the record of such organizations gives us the right to believe them capable of any degree of deviousness or subterfuge. But if the accounts we have are reliable, the MIB perform feats that are beyond the powers of even the most lavishly funded government security agency. The ability of Bender's visitors to manifest in his room, and similar episodes in other reports, all testify to paranormal powers that demand a non-human explanation, unless we reject them as pure fantasy. Similarly, the motivation and methods of some of the MIB are hard to explain in terms of government agencies. On 1 March 1967 a United States Air Force official memo was circulated under the heading 'Impersonation of Air Force Officers', warning that such impersonations were taking place: in the bluff and counter-bluff world of the security man, this can be interpreted in many different ways; but at least it acknowledges the public character of the phenomenon. And probably the Air Force was being honest in dissociating itself from the MIB: certainly the secret agent hypothesis seems inadequate to account for the bizarre case that follows.

In September 1976 Dr Herbert Hopkins, a 58-year-old doctor and hypnotist, was acting as consultant on an alleged UFO teleportation case in Maine, USA. One evening, when his wife and children had gone out to a movie, leaving him alone, the telephone rang. The caller identified himself as the vice-president of the New Jersey UFO Research Organization, and asked if he might visit Hopkins to discuss the case.

Hopkins agreed. He said later that it had seemed at the time the logical thing to do, just as Richardson in the previous case had willingly answered questions from complete strangers. He went to his back door to switch on the light so that his visitor could find his way from the parking lot, and saw the man already climbing the porch steps. 'I saw no car, and even if he did have a car, he could not have possibly gotten to my house so quickly from any phone,' he later reported.

But at the time Dr Hopkins felt no particular surprise as he let his visitor into the house. The man was dressed in a black suit, with black hat, tie and shoes, and a white shirt. 'I thought, he looks like an undertaker.' His clothes were immaculate, suit unwrinkled, trousers sharply creased. When he took off his hat he revealed himself as completely hairless, not only bald on top but also without eyebrows or eyelashes. His skin was dead white, his lips bright red; in the course of their subsequent conversation he brushed his lips with his grey suede gloves, and the doctor was astonished to see that his lips were smeared and the gloves stained with lipstick.

It was only later, though, that Dr Hopkins gave any thought to the strangeness of his visitor's appearance and behaviour. At the time he sat discussing the case in a normal manner. When he had given his account, his visitor said that Hopkins had two coins in his pocket, which happened to be the case. He asked the doctor to put one of the coins in his hand: unquestioningly, Hopkins did so. The stranger asked him to keep his eyes on the coin: as he watched, the coin seemed to go out of focus, then gradually vanished. 'Neither you nor anyone else on this plane will ever see that coin again,' his visitor told him.

After talking a little while longer on UFO topics, Dr Hopkins noticed that his visitor's speech was slowing down. The man rose unsteadily to his feet and said, very slowly, 'My energy is running low — must go now — goodbye.' He walked falteringly to the door, and went down the outside steps one at a time, unsteadily. Hopkins saw a bright light shining in the driveway, bluish-white and distinctly brighter than a normal car headlamp: at the time, however, he did not think otherwise than that it was his visitor's car, though he neither saw nor heard the car itself.   •

Later, when his family had returned, they found marks that could not easily have been made by a car because they were in the middle of the driveway where the wheels would not have run. By next day, though the driveway had not been used in the meantime, the marks had gone.

Dr Hopkins was much shaken by the visit, and willingly complied with his visitor's instruction — for which no plausible reason was given — that he should erase the tapes of the hypnotic sessions he was conducting in connection with the UFO case. The readiness of Hopkins to go along with this totally unauthorized demand is testimony to his sense of the reality of the incident. What makes the episode more extraordinary is that he was able to establish that no such organization exists as the New Jersey UFO Research Organization: whatever he was, the man was an impostor, though this clearly made him no less imposing in the doctor's eyes, perhaps even the reverse.

A few days after the visit, Dr Hopkins' daughter-in-law Maureen was telephoned by a stranger who claimed to know her husband John, and who asked if he and a companion could visit them. John, who did not recollect meeting them before, met the pair at a nearby restaurant and brought them home: as with Dr Hopkins, the percipients seem to have performed quite unusual actions without questioning them at the time. The stranger and his female companion were both seemingly in their mid-thirties, and wore curiously old-fashioned clothes. The woman looked particularly odd: her breasts were set very low, and when she stood up it was as though there was something peculiar about the way her legs joined her hips. Both the visitors walked with very short steps, leaning forward as though afraid of falling.

They accepted drinks, but did not so much as taste them. They sat awkwardly together on a sofa while the man asked a number of detailed and often intimate questions, about their television and viewing habits, what John and Maureen talked about. All the while the man was pawing and fondling his female companion, asking John if this was all right and whether he was doing it correctly. While John was out of the room the man asked Maureen if she had any nude photographs of herself.

When, ultimately, the strange couple announced that it was time for them to leave, there was a curious difficulty. The man stood up but made no move towards the door. He was between his companion and the door, and the woman seemed unable to walk round him. Eventually she turned to John and asked 'Please move him, I can't move him myself.' Then, abruptly, the man left, followed by the woman: both walked in straight lines, and neither said anything more. [44e]

It is tempting to dismiss this bizarre incident as pure fantasy: but there are several people involved, and the detailed observations offer, despite their surreal character, an internal consistency that is curiously compelling. It points towards a scenario in which some kind of alien — perhaps robotic — creature is trying to adapt to human behaviour, but is having difficulty in translating its knowledge of what to do into action. It is, indeed, a classic science fiction situation. For this reason, such accounts tend to be explained sometimes as projections from the percipient's imagination, sometimes as 'displays' presented by the aliens with the deliberate intention of misleading the percipient.

It would be idle to pretend that the evidence for the Men in Black is anything other than flimsy: but it is precisely the ability of the myth to establish itself on so weak a foundation that makes it especially relevant to our study. Here we have at least thirty cases in which there is some reason to believe that something anomalous occurred, involving an entity or entities that can be loosely described as 'men in black'; and we see further that this type of case has developed its own characteristics, making it distinct from any other of the kinds of apparitions we have so far considered. Even if we discount the more sensational phenomena, we are left with recurrent reports of entities that offer the following features:

★ They have access to information which is, if not totally inaccessible to the general public, then obtainable only by very devious means necessitating a highly sophisticated explanation. We do not have to premise clairvoyance or any other kind of extra-sensory perception to explain this, though it is admittedly a characteristic that is shared by some kinds of apparition. But if we do not suppose unusual powers of one kind, we must find unusual powers of another!

★ They behave in ways that suggest a non-human origin. They appear and disappear in a non-veridical way, which again is a characteristic they share with apparitions, visions and so on.

★ Their actions occasionally involve teleportation, and they are described as communicating by extra-sensory means.

★ Their behaviour, though superficially logical, does not on analysis seem rational or consistent. Not only are they not who they claim to be, but they do not appear to have the backing they claim to have: though they frequently threaten their 'victims', there is no record of any serious injury or even physical assault. Some witnesses have claimed murder attempts — cars which seemed to be trying to run them down, bullets which just missed them — but this may be no more than paranoia. (It could be argued that perhaps there were other, successful, attempts: but not only is there no evidence for this, there are also no reports of, say, people being run down or shot at and *surviving*.)

The most probable explanation for the Man in Black phenomenon is that it is a hallucination, projected as a result of the percipient's private fears, and given a specific shape and form based on prevalent notions of the CIA and other such agencies. But such an explanation presupposes the necessary psychological mechanism, whose existence has yet to be demonstrated. And it does not account very convincingly for the uniquely lifelike nature of the MIB entity. It is true that many of the reports we have of other types of entity describe them as being more *vivid* and *real* than anything the percipient has ever known in real life: but no other is described as being so *lifelike*, so completely integrated into the context of real life as to convince the percipient that he has not experienced a hallucination, whether at the time or subsequently. MIB percipients do not simply have a convincing sight of their entities: they engage in a complete and consistent series of actions — they open the door for them, talk to them, offer them drinks, and so on, often over an extended period of time.

We are justified, therefore, in treating the MIB as a specific category of entity in its own right, and in demanding of any field explanation that it account for the MIB or, alternatively, give us good reasons for regarding it as outside our field of study.

1.11 UFO-RELATED ENTITIES

UFO-related entities, like religious visions, are associated with a specific belief-system: but whereas religious visions tend to be seen by percipients who are already believers in the system — which generally means that they are devout Catholics — this is not the case with UFO-entities. These are generally seen by people who do not, so far as we can see, have any predisposition to see UFOs or even to believe in their existence.

Over the past three decades, reports of UFO-related entities have been made on a scale which makes them by far the most commonly reported type of paranormal entity outside dreams. Yet they present us with an extraordinary paradox, which is that they are an anomalous phenomenon associated with a phenomenon which is itself anomalous, to such a degree that its existence, let alone its nature, is still very much in doubt.

Unidentified Flying Objects continue to be the most elusive of phenomena, with the result that they are not taken seriously by more than a small minority of scientists. A good many amateur investigators, many of them working to high standards, are active in the field, and as a result of their efforts a very great deal of information has been collected: but the net result of this information is simply to underline the extraordinarily complex nature of the problem. One by one, each simple solution to the UFO challenge has been shown to be inadequate: we are now faced with a phenomenon whose explanation, whatever it is, *must* result in altering our knowledge of the universe and its laws. And this is true whether or not the UFO is found to have any physical basis; whether or not it is determined to be a psychological construct; whether or not it is of extraterrestrial origin; whether it is organic or artefact. [40]

No assumption is made in this study as to which if any of these interpretations of the UFO is valid: we are not concerned with the UFO as such. But the entities that have so frequently been reported in connection with UFO sightings do very definitely qualify for our attention. They give many indications of sharing the attributes of some of our other categories, notably visions on the one hand and folklore entities on the other, while in their behaviour they display many of the characteristics associated with ghosts and hauntings. At the same time, they have features of their own that distinguish them from all other categories.

The very great number of UFO-entity reports is extremely fortunate for our study, for it provides us with a generous collection of recent and accessible data on which to draw, and many cases have been studied with a thoroughness rare in other fields: for instance, there are relatively few apparition cases in which witness statements have been tape-recorded, in which lie-detector procedures have been carried out, in which the witness has given his testimony under hypnosis, monitored by video-cameras for subsequent analysis. None of the other anomalous phenomena in human history has been studied with as much care; and though much UFO literature is naive to the point of lunacy, much too is of a very high standard. Moreover, the fact that all the material is more or less contemporary means that it can be checked in a way which is not true of any other category. As we shall see, a small but growing number of behavioural scientists have appreciated the unique opportunity provided by the UFO for conducting research into the sociological and psychological parameters of entity reports. Already the results have presented findings that may well prove to be among the most illuminating revelations of human behaviour currently being added to our knowledge.

By UFO-related entities we mean entity-sightings that occur in close association with UFO-sightings, or which give reason, even though no UFO is actually seen, for thinking that they are so associated — for example, the entities may tell the percipient that they come from a UFO, or they may be seen at the same time that a UFO is reported independently in the neighbourhood. Sometimes the association is only by inference: thus, if a percipient receives a visit from an alien humanoid who gives every indication of being from elsewhere than our own world, it may be considered reasonable to class him as UFO-related, even though he may have found some other means of travel than an alien spaceship to get from there to here!

The spectrum of entities ranges from creatures who seem quite specifically to be the crews or passengers of UFOs which themselves seem to be alien spacecraft — that is, creatures who are the precise equivalent of our own astronauts — to 'cosmic brothers' who present themselves as representatives of some distant civilization, which may even exist on some other plane of reality. If they are, in fact, physical beings, extra-terrestrial visitors or suchlike, then of course they would fall outside the field of our inquiry, just as would visions of the Virgin if it should turn out that it is Mary herself in her physical body who is responsible for the sightings. But this is not an assumption we can make in either case, as the following case histories demonstrate. These are three 'classic' accounts, chosen because they have been investigated very thoroughly on both the physical and the psychological levels: and while none of them can be said to be 'solved', it is at least evident that each presents a serious case of — *something*.

Valensole, southern France, 1 July 1965

Arriving early in the morning to work in his lavender fields, 41-year-old Maurice Masse heard a curious whistling sound, which on investigation he found to come from a machine shaped like a rugby football, the size of a small car, standing in one of his fields. Two entities, as tall as eight-year-old boys, were inspecting his lavender plants. He approached them stealthily, believing they were damaging or stealing his plants, then realized they were not ordinary people. They became aware of his presence, turned, and pointed a pencil-like device at him which held him as if paralyzed, unable to move but not unconscious. They then returned to their craft and Masse saw it take off and float silently away; it left traces in the field which investigators were able to examine, and plant growth was affected for more than a year.

The entities were described by Masse as about 1.25 metres tall, wearing close-fitting grey-green clothing, no head covering, with heads like pumpkins, high fleshy cheeks, lipless mouths, very pointed chins, and eyes which slanted away. As with a high proportion of UFO-entities, the physical description is of near-human creatures, seemingly adapted to physical conditions on Earth, yet with sufficient difference from humans to indicate an extraterrestrial origin.

Masse had a good reputation in his community, and though the case was extensively investigated by the police and by ufologists, no indication has been found of any deception. On the other hand, apart from ambiguous physical traces and an inconclusive confirmatory sighting by a neighbour, the report consists of a single-witness visual sighting, like the majority of entity cases. [44 a]

Ashland, Nebraska, USA, 3 December 1967

About 2.30 a.m., police patrolman Herbert Schirmer, alone in his patrol car, approached an intersection on the outskirts of Ashland, and saw, at a distance of about half a kilometre, a number of lights which looked as though they came from the portholes of a craft standing on the ground: it was shaped like a football, with what seemed to be a catwalk running round it and tripod legs underneath. As Schirmer watched, the machine rose into the air with red-orange flames emitted from its underside and making a siren-like sound. He returned to the police station, to find that the time was now 3 a.m., much later than seemed accounted for by his actions. He logged the sighting, half-seriously, as a flying saucer.

Invited to co-operate in further investigation, Schirmer agreed to submit to time-regression hypnosis. In the hypnotic state he was allegedly able to recall what had happened during the 'missing' twenty minutes, providing information which he had not been able to con-sciously recall:

The craft actually pulled me and the car up the hill, towards it; and then as I was going up the road, and the car came to a dead stop, a form came out from underneath the craft and started moving toward the car . . . another one was coming out . . . the one that was standing in front of the patrol car had sort of a boxlike thing in his hand and kind of flashed green all around the whole car. And then one approached the car and reached in and touched me on my neck, at which I felt a sharp amount of pain — and then sort of stood back and sort of moved his hand and I just came right up out of the car, standing right in front of him, and he asked me 'Are you the watchman of this town?' and I said 'Yes, I am.' He said 'Watchman, come with me,' and we went up into the craft; and as we got to the craft, he took me up into what I call the first level of it . . . and we're standing there looking at these, like 55-gallon barrel drums in a big circle, and had black cables being connected to each one of them. And then right in the centre of the room, as we looked up, was like a half of a cocoon, and it was spinning, giving off bright colours like the rainbow. And he said 'Watchman, this is our power source — reversible electrical magnetism . . . The reason we're here is to get electricity,' and they extracted electricity from one of the power poles there, which led to the main power source there in Ashland.

Schirmer was taken on a tour of inspection, shown various sights, then returned to his car, at which point his host said 'Watchman, what you have seen and what you have heard, you will not remember. The only thing that you'll remember is that you've seen something land and something take off.'

He described the entities as having very high foreheads and long noses, slit-like mouths and sunken eyes, round with cat-like pupils. Their complexions were greyish-pink, and they were dressed in uniforms.

Like M. Masse, Herbert Schirmer was well known and respected in his community, and there was no reason for doubting his word; but like the French case, his rests entirely on his own word. Nobody else saw the object, it left no traces or other confirmation. The difference between the two cases is, of course, that M. Masse consciously recalled his encounter, and never doubted its physical reality; whereas Schirmer's conscious memory was only of a curious sighting, with no question of entities until he was invited to go through the experience while under hypnosis. We shall be looking more closely at what is involved in hypnotically recalled experiences in section 2.1: but we must not lose sight of the fact that Schirmer's case is one of many dozens, and that together they present a type of experience which constitutes a new and very puzzling phenomenon. There can be no question that Schirmer did have a real and unaccountable experience that night: no policeman in his right mind — and Schirmer was examined by competent psychologists and considered to be sincere and reliable — would enter in his official report an experience whose reality he had any cause to doubt. [10]

Desert Center, California, USA, 20 November 1952, and later.

George Adamski's encounter with an entity naming itself Orthon, and who claimed to originate from the planet Venus, was the first widely publicized case in which a human claimed direct contact with an alien being visiting Earth. Adamski had had earlier UFO sightings, and had come to believe that by going to the right place at the right time he might be able to experience a more direct encounter. A self-appointed 'professor' of a mystical philosophical system, he considered that he might be favoured with some personal communication appropriate to his standing. This hunch proved well-founded when he and some friends drove out into the desert and saw a small saucer-shape UFO emerge from a larger cigar-shape 'mother ship'. It came down to ground level, and an entity emerged with whom Adamski entered into 'conversation' by means of telepathy. Orthon told him that the Space People whom he represented had friendly intentions towards Earth, but were concerned about 'radiations from our nuclear tests', and advocated a philosophy of life very similar to that which Adamski himself had been preaching for many years — this, of course, was why he had been chosen by them for this meeting.

The entity was described by Adamski as a smooth-skinned, beardless figure, wearing clothing like a ski suit with a broad belt round the waist: he wore shoulder-length blond hair. On subsequent occasions Adamski met other visitors, from Venus and other planets, learning from them that all other worlds except Earth were co-operating in peaceful constructive progress. He was taken on several space flights, during which he saw the cities on the far side of the moon and other wonders, besides meeting many aliens and finding that they are in many respects like Earthpeople but more beautiful, more friendly, wiser and of course technically more sophisticated.

The fact that Adamski was accompanied during his first encounter — though his companions remained at a respectful distance of more than a kilometre — means that there is some confirmation of his sighting, though, coming from loyal friends and disciples, it is not quite the independent confirmation one would prefer to have. [93]

There is much more to the Adamski case than this encounter, which was the first 'contactee' case to be widely publicized. In the footsteps of Adamski many others announced that they had had comparable experiences: in not a single instance was there convincing supporting evidence, and virtually every one comprised a single-witness statement of varying degrees of implausibility. No serious ufologist is prepared to take them at their face value, but they represent fascinating case material for the behavioural scientist, not least because many of them attracted cults of followers who had not been privileged to share the original experience, but who enthusiastically hung on the words of the visionary. The parallel with the religious visionaries is, in this respect at least, very striking.

Unlike Masse and Schirmer, Adamski gives us good reason to doubt his word. His descriptions of his space journeys are full of details which — now that we have embarked on space exploration for ourselves — are demonstrably false, and this has encouraged many researchers to dismiss him as an out-and-out charlatan. The truth is probably more complex, and I hope that our inquiry will lead us closer to that truth.

Certainly we have no right to do what many ufologists did when such cases as these were first reported — to dismiss them as simple imaginative fantasy. Whatever they are, they are not simple. Here are some of the complicating factors:

★ There are a very great many accounts, given by people who in many cases cannot by any reasonable means have been aware of the experiences of others, that offer similar features, not just in broad outline but in tiny details, such as the physiological characteristics and uniform worn by the entities, their manner of movement, and so on. The little beings reported by Masse, for example, are very similar to those reported by many other percipients: if we are to reject the obvious explanation — that they were all seeing the same kind of entity — then we have to establish how it comes about that people like Masse, apparently uninformed about and uninterested in these matters, should all give such similar descriptions.

★ The great majority of UFO-entity percipients impress investigators with their patent sincerity, and this is confirmed when formal psychological testing has been carried out. Dr Leo Sprinkle, of the University of Wyoming, reporting to the Condon Committee on the Schirmer case, concluded that 'in my opinion, the events described by Sgt. Schirmer are "true" in his experience', while admitting that 'the present evidence does not answer the questions regarding the source, method, and purpose of communicating the additional information.'[19]

★ There is some evidence that in some of the cases there was an initial event that was genuine, whatever the alleged entity encounter that followed may have been. The most remarkable instance of this is the case of Barney and Betty Hill[48] in which a simple UFO sighting developed into a closer encounter in which occupants were observed; and under hypnosis the two witnesses alleged a further stage in which they were abducted on board a UFO for physical examination. Few researchers have felt justified in rejecting all parts of the story, but many are disinclined to accept all as equally true: it could well be that an incident on one level led to an event on another.[39] This necessarily involves a somewhat elaborate explanation.

★ The circumstances of an alleged encounter, if there was no encounter in 'real' terms, must somehow be accounted for. For example, in December 1976 an elderly Frenchwoman was returning home by car alone at night, when she had — so she claimed — an abduction experience on board a UFO. If her story is not true, then either her car was parked at the side of a very busy road and nobody saw it; or she went elsewhere for the period of time of the encounter, then returned to the scene of the incident to continue her journey. [125] Such down-to-earth difficulties occur in many such cases, and have led to the recognition of two awkward aspects of encounter cases, the 'missing time syndrome' and the 'traffic-free syndrome'. Both have their parallels in other types of entity experience, of course: rarely, if ever, is an entity-sighting interrupted by mundane events on the terrestrial plane!

A considerable part of the next section of this study is devoted to parallel findings relevant to UFO-entity sightings. The intensity of investigation of these phenomena has resulted in a body of experimental material which throws light not only on this category of sighting but on others also. These findings make it clear that, whether or not the UFOs and their occupants are extraterrestrial visitors, the experience of seeing them has much in common with other apparition experiences. The following case is very revealing in this respect:

Zimbabwe, Africa, 31 May 1974

A young couple, 'Peter' and 'Frances', were driving at night on an open country road, when they saw an anomalous light that came from an object in the air above them. At the same time it seemed to Peter, who was driving, that control over the car had been taken away from him. The car grew very cold, strange signals came over the radio: subsequently they found that they had travelled the distance at a seemingly impossibly high speed, and used less than two litres of fuel to cover 288 kilometres. They reported their experience to UFO investigators, who employed hypnotic regression to try and elicit further details.

The hypnosis was most successful, and a wealth of information was forthcoming as to how the spacecraft had taken over control of the car. The customary explanation was given, that the spacecraft came from a highly developed civilization in the 'outer galaxies' — not specifically identified, of course — whose inhabitants were concerned about what was happening on Earth. (We shall look more closely in a later chapter at this tendency towards vague stereotypes.) Particularly relevant to our study is this extract from 23-year-old Peter's recorded statement while under hypnosis:

We were programmed inside the motor car . . . My wife fell asleep, or was put to sleep by the radio which was the voice of 'them', so she can't remember very much of what happened inside the car. A form was beamed straight to the back seat of the car and sat there the entire journey and told me I would see what I wanted to see. If I wanted it to look like a duck, then it would look like a duck; if I wanted it to look like a monster, then it would look like a monster.

When asked if the beings had, nevertheless, any 'real' form, Peter replied, 'They are physical beings. The one that was in the car and those in the spaceship were all identical in size, colour, looks, shape, weight. The same basic form as humans — large trunks, necks, hairless, two arms, two legs, no toes. No reproductive organs at all.'[41,44c]

Hypnotic regression is notoriously untrustworthy, as we shall see in a later chapter; but in this case the transcript permits us to hypothesize that the entire episode was nothing more than Peter's own fantasy, generated in the course of an altered state of consciousness resulting from driving long distances on featureless roads at night. The fact that Frances had dropped off to sleep — whether induced by spacebeings manipulating the radio, or simply because she was tired — means that Peter's experience was to all intents and purposes unshared and unconfirmed: Frances does confirm the original sighting, and the factual 'waking' details such as the extraordinary fuel consumption, but none of the rest.

Significantly, Peter revealed that this was not his first UFO experience. He explained the 1974 event (in which, as other parts of the transcript describe, he claims to have been abducted on board the spaceship while the spacebeings controlled the car) by referring to a very brief contact at the age of thirteen, ten years previously, though this had been a very minor incident, in which no entities were observed.

Another significant statement was to the effect that these spacebeings, whom he describes as 'very clever', 'took about seven seconds to find that I had communicated with the past, and had control over my mind to be able to give myself post-hypnotic suggestions'. This tells us that Peter was by no means a typical witness: the number of motorists in Zimbabwe skilled in self-hypnosis and astral projection must be very few. Besides, his statement that the spacebeings took 'seven seconds' to learn that he had this capability seems curiously precise.

So here we have an unusual person travelling in conditions that are conducive to fantasizing. An unexplained object is seen, and this triggers off a fantasy along lines which bring together various elements from Peter's interests, concerns and experience — is that it? But then, what about the 'missing time' element — the impossible speed of their journey; and what of the equally impossible fuel consumption? Was the first an illusion, and did Peter in fact stop and refuel while in a trance state, having no recollection of it even when hypnotized? Complexities like these make it impossible for us to relegate even an apparently 'personal fantasy' case to a purely psychological category.

Several attempts have been made to classify the varieties of alien entity reported by witnesses. The classifiers do not of course believe themselves to be laying the basis of an extraterrestrial anthropology; they are simply trying to establish what common elements recur. There is a general consensus that, though hardly two entities are precisely alike, they fall into four main divisions:

★ Totally human-like, as for example those encountered by Adamski.

★ More or less human, but with slight differences — like Peter's, which were like us but lacked toes and reproductive organs.

★ Definitely not human, but humanoid, seemingly well adapted to the physical conditions of our Earth, but with notable differences such as over-large heads, wrap-round eyes and so forth. M. Masse's aliens are representative of this category.

★ Animal-like, with a tendency towards such species as bears or monkeys, but seemingly more intelligent. These beings are generally seen only in purely visual cases, where there is little if any interaction, and no communication.

Almost without exception, UFO-related entities are bi-pedal, and have heads, trunk and limbs disposed much as ours are; their faces have much the same features, disposed in much the same configuration — single eyes and no mouths, though occasionally reported, are rare. In their behaviour, too, the entities are not markedly different from humans; but just as their vehicles are able to out-perform anything we on Earth can put into the skies, so their occupants have abilities we humans cannot match, such as being able to glide up off the ground into their ships, make use of extra-sensory communication, and so on. But this only makes what they actually *do* — or do not do — the more bewildering. For the motivations they profess are naive in the extreme; and their actions — inspecting plants, taking soil samples and so forth — are so crude when compared with the highly sophisticated efforts of our own space explorers as to be quite unconvincing. They profess themselves benevolent, and they have given us no reason to disbelieve them: at the same time they are seldom positively friendly, though there are a few cases in which the entities have told the percipient that he has been specially chosen, and have given him privileges such as showing him round their ships or taking him on journeys in space.

Much has been written about the dangers of extrapolating from human experience: we should not, it is said, expect alien visitors to behave as *we* would if we were visiting *their* worlds. Nevertheless, one feels that one should be able to detect some kind of logic in their behaviour even though such logic was not ours nor even comprehensible to us: by analogy, there are many species of animal that indulge in behaviour patterns which we humans cannot understand, but at the same time we can see that they must have a meaning for the creature concerned. With UFO entities it is not possible to give them the benefit of such a doubt: their conduct is almost always bafflingly meaningless to our eyes. Take, for example, their often professed desire to warn mankind of the dangers of continuing along our present self-destructive road with regard to nuclear weapons and so forth. How do they go about delivering these warnings? They arrive stealthily, generally in a remote location and at night; more often than not they appear only to a single individual, who is never anyone in any kind of influential position. To this individual they give a verbal message, vague in its terms though often very lengthy, which he is to pass on to his fellow creatures. It contains no specific details, no instructions, no verifiable facts, nothing that indicates superhuman knowledge; nor is it supported by any proof of identity, no authorization from some superior power, no evidence that the message emanates from a source that has the right to dictate to mankind . . . The parallel with the messages given at Fatima and Garabandal, in spiritualist seances and to dozens of 'automatic writers', is very evident.

There is one major difference between UFO-related entities and any of the other categories in our study: they are susceptible of an entirely physical and material explanation. They may indeed actually be the crews and passengers of alien spaceships. The temptation to interpret them in this way has been, for perhaps a majority of those who study the UFO phenomenon, irresistible. But while there is much in the reports to support this interpretation, there is much too that speaks against it, and the same flexibility of interpretation that is necessary if we opt for the real-life-aliens explanation can be used to place them in a totally psychological context, where all the inconsistencies and illogicalities can be ascribed to defective fantasizing on the percipient's part.

A characteristic aspect that links some UFO-related cases to some visions is the 'chosen' element: some witnesses, like Masse and Schirmer, regarded their experiences as accidental — if you or I had been there at the time, we too would have had the experience; whereas others, including Adamski and 'Peter', believed they had been selected for the experience. This is a characteristic shared with visionary cases like that of Lourdes and Fatima: just as we find it hard to understand why the Virgin Mary should select Bernadette Soubirous out of so many millions of potential visionaries, so we ask Why George Adamski? It is a question that we do not have to ask if we suppose that the selection came from the percipient, rather than from the entity.

Are we to suppose that, subconsciously, all the witnesses — Masse and Schirmer no less than Adamski and Peter — were unconsciously 'seeking' their encounter? And in that case do we have to suppose that every UFO-percipient is also responding to some subconscious motivation? Or should we look on the contactees as extreme cases along a spectrum that is for the most part made up of experiences that fall within the accepted limits of normality? This is how Dr Berthold Schwarz views the contactee syndrome from the psychic/psychiatric angle:

> A contact is not just an isolated event in an individual's life but something that must be viewed in the larger context of his past history and his postcontact experiences, attitudes and behaviour . . . Many have dissociative personalities, in some cases even multiple personalities. They are susceptible to trance states . . . Yet they manage to lead normal, responsible lives, holding down jobs, heading families, refraining from antisocial conduct. But often that changes when they have their UFO sighting — they blow up like an erupting volcano. Did their psychological problems cause them to imagine the experience — or did a real experience cause the problems to surface?
>
> We simply don't know. We do know that after this supposed experience percipients may undergo alternating states of consciousness, slipping in and out of trance states, during which they may channel messages from oddly named entities. So far as content is concerned, these messages are pure garbage. Still, whatever their cause, whatever their source, they 'come through'. Another thing that happens is the unleashing of psi phenomena around the percipient. Perhaps it's to be expected, since trance-like states are conducive to the production of ESP and psychokinesis.
>
> Maybe the UFO experience is a way for these people to fulfil themselves. Sometimes it turns out that the UFO contact successfully serves the percipient's need; other times it turns out that it does not and the person ends up even worse off than he was before. [138]

It would be a bold theorist who would assert that all UFO-related entity-sightings are entirely subjective. The stimulus for the experience, as Schwarz accepts, may be 'any one of a number of things, from a real "objective" encounter to some sort of psychic projection to pure hallucination'. But what of the entity-sighting? Can we draw a clear line between the contactee cases, in which the psychological element certainly seems paramount, and those that give every appearance of having been objective? Can we put the Masse and Schirmer cases, say, on one side, and the 'Peter' and Adamski cases on the other? At the present stage of our inquiry, at least, we would hardly be justified in doing so.

Many of the questions raised in this brief notice of the UFO-related entity will be considered again later in the light of further evidence. For the time being, let us simply face the fact that in the UFO-related entity we have a separate and specific category that cannot easily be accommodated in the same pigeonhole as any of the others we have studied. UFO entities look and to a large extent behave as though they are as solid as human beings; they have longer and more detailed communication with their percipients than almost any other entity; they are more clearly associated with a particular cultural context than any other category apart from religious visions; and they are readily identified by percipients as being what they seem to be, in the same strange way that religious visions are for the most part known to be so and not something else. In short, so far as their percipients are concerned, UFO-related entities form a category in their own right, and for the time being that is how we too must treat them.

1.12   ENTITIES AS FOLKLORE

When, in 1959, Carl Jung wrote a book about UFOs, he sub-titled it 'a modern myth of things seen in the skies'. Without denying the possibility that UFOs may possess some degree of physical reality, he nonetheless saw the phenomenon as a living example of myth in the making. Similarly, earlier in our study we considered the possibility that in the current accounts of 'Men In Black' we have another opportunity to watch folklore being created before our very eyes.

A further dramatic instance of this process may be the 'Phantom Hitchhiker', an entity who recurs in many reports, of which the following is typical:

In January 1976 two young men from Bagnères, France, are trying out a brand new car. On the outskirts of town they pick up a young woman hitchhiker: because the car is a two-door model, the passenger has to get out and hold his seat forward so that the girl can get in the back. They drive on, and the driver, keen to test the performance of the new machine, drives steadily faster, until the hitchhiker warns 'Look out, fellows, there've been a lot of accidents in the bends you're just coming to — I know all about them.' The young men turn their heads to reassure their passenger, and slow down to oblige her. After negotiating the turns successfully, one of them remarks, 'You see, Madame, other people may kill themselves here, but we got by alright.' Their passenger does not reply, whereupon the young men turn — and see that the back seat is empty. They stop the car, to make quite sure; but there is no question of it: the hitchhiker has gone. [83]

The story has been told in many forms, set in many different parts of the world. Sometimes it is embellished with the subsequent discovery that the young lady had herself been killed in precisely that spot some time earlier.

The obvious explanation is that the young men saw the ghost of the victim, who had returned to Earth from a sense of duty, to warn others travelling the dangerous stretch of road. But while this is the simplest and most logical explanation, it is not the only one, and is of course open to some objections. Apart from the fact that the notion of *any* person returning from the dead to give warnings to the living is far from having been proved as fact, there is the wide prevalence of this specific type of case. It is not easy to believe that so many victims of precisely this type of accident should feel the urge to return to warn living people — and total strangers at that — while the victims of other kinds of accident do not seem to feel a similar compulsion.

But what are our alternatives? We can construct a model based on extra-sensory perception: we may suppose that the driver had a clairvoyant awareness of the danger ahead of him, and that his subconscious mind, anxious to present the warning in as graphic a form as possible, presents it as though it emanated from an external source, lending force to what otherwise would have been just a vague premonition that might have been ignored. But here we have also to account for the fact that the driver's companion also saw the girl, to the extent of getting out of the car and holding the seat forward so that the ghost could get in the back seat: this is a more elaborate illusion than all but a few rare stories of apparitions, yet we are asked to believe that it has occurred on many occasions!

There is an additional complication in the case of those versions of the story in which the warning entity is subsequently identified as a real-life victim. We can suppose that the subconscious mind, creating its scenario, has made use of a real person instead of creating an imaginary one: but it does require that an additional element of extra-sensory perception — this time of clairvoyance or telepathy — is involved.

True to our principle of not rejecting any explanation no matter how remote, until we have considered all the evidence, we can perhaps entertain such an explanation for the time being, remembering that the ability of the subconscious mind to create fantasy scenarios has no known limits. But we must recognize that the ESP explanation begs as many questions as the returning-spirit explanation. For if it is not a proven fact that people return from the dead to warn the living, no more is it proven that the subconscious mind can and does make use of clairvoyantly acquired material in this way.

It is hardly surprising, therefore, that most students of the phantom hitchhiker phenomenon — and its widespread character has meant that a good many researchers have given it their attention — have tended to relegate it to the category of folklore. But this is no explanation whatever, for the nature of folklore is itself a much-debated question. One view of it, for example, is that folk tales are essentially fiction, but that the narrators — or at any rate the original narrator — gave them a local habitation and a name in the interests of greater verisimilitude and thereby greater impact on their listeners. An opposite view is that they start as real-life events in a specific place, but are relocated, for dramatic reasons, by later narrators.

So while we need not seriously consider the possibility that every one of the accounts is true, we do have to consider the possibility that *one* of them is true, and that all subsequent versions are re-tellings of the original story, whether conscious or unconscious. This seems neat and plausible, were it not for the fact that more than one researcher has found substantial evidence of the genuineness of *later* stories. Thus Michael Goss, author of a book-length study of the subject[57], felt compelled to say of one case he had personally investigated: 'I find the Stanbridge case convincing, not as a piece of relocated folklore but as a paranormal fact.'[56] It may be significant that the case he found so convincing was one in which there were no words spoken: we might conclude from this that the more elaborate the tale, the more circumstantial the detail, the more we may suppose it to be an elaboration of the original, dressed up by the narrator for purposes of effect. The true original, the archetype, may be the simplest, the least embellished with 'convincing' detail.

But a recent episode in the phantom hitchhiker saga suggests otherwise. On 20 May 1981, four young people aged between twenty-five and seventeen were returning from the beach to their home town of Montpellier in southern France. It was some time after 11 p.m. The two young men were in front, their girlfriends behind. Quite soon after starting the journey, they saw a woman standing by the roadside, apparently thumbing a lift: she was dressed in white and wore a headscarf. Though he did not usually stop for hitchhikers, and though the car was full, Lionel, the driver, said that they could not let a woman stand alone on the roadside at that hour of night. He stopped beside the woman and told her they were headed for Montpellier: she said nothing, but nodded her head. Then, because it was a two-door car, the other young man got out, and one of the girls, so that the woman could get in: she sat between the two girls in the middle of the back seat, and could now be seen to be a woman in her fifties.

Not a word had been spoken so far, but suddenly, after travelling some distance, the passenger cried out, so loudly as to drown the car radio, 'Look out for the turns, look out for the turns!' She gesticulated with her right hand and added 'You're risking death'.

All of them gazed intently at the road as if an imminent danger really did threaten: Lionel slowed down as he negotiated the bend. Then the two girls screamed: their fellow-passenger had abruptly vanished.

Although it was impossible that she could have got out of the car, they had a look round the area but of course found nothing. They then drove immediately to the police station at Montpellier and told what had happened. Sceptical at first, the police were convinced by the evident fear and sincerity of the four young people, and though unable to find any explanation, had to accept that a real incident of some kind had taken place.[84]

The account is archetypal, and if it were not so well documented we should have no hesitation in relegating it to the category of old wives' tales, told by 'a friend of a friend' who swears he knew the person it happened to . . . But in this case we have four witnesses, whose stories all agreed, who went voluntarily to the police immediately after the incident: it is hard to believe they were playing a practical joke. But if not, if their story really occurred, then it has all the appearance of life imitating legend!

1.   Glenda's own drawing of the 'spacewoman' who appeared in her bedroom in 1976 (see page 15).

2. Entities seen in dreams only rarely take the form of malevolent beings, but these are the most notorious. This nightmare, depicted by Langlumé in an 1840s lithograph, could be the product of an uneasy conscience — or of over-eating.

3. The doppelgänger, rare in real life, is frequent in literature. In Edgar Allan Poe's story, William Wilson is haunted by his own double, depicted here by Harry Clarke.

4a. The fact that apparitions usually appear perfectly normal to the witness makes them difficult for artists to depict. The housemaid whom Miss Leigh Hunt followed upstairs in Hyde Park Place in 1884 (see page 79) did not really look as 'ghostly' as this, but was taken for a real person until she vanished at the top of the stairs. From the *Strand*, 1908.

4b. Hauntings are usually explained as memory-traces — 'imprints' left in a place by someone, now dead, who used to live there. It is attractive to suppose that this is the explanation for photographs such as this, taken by Mr Bootman, a bank manager, in the empty church at Eastry in 1956. (*Photograph courtesy of Andrew Green.*)

5. The so-called 'Angels' alleged to have been seen during the Battle of Mons, on 23 August 1914, were more generally reported as bowmen and identified with the English infantry of the Hundred Years War (see page 114). The scene was imaginatively depicted by Forrestier in 1915 for the *Illustrated London News*.

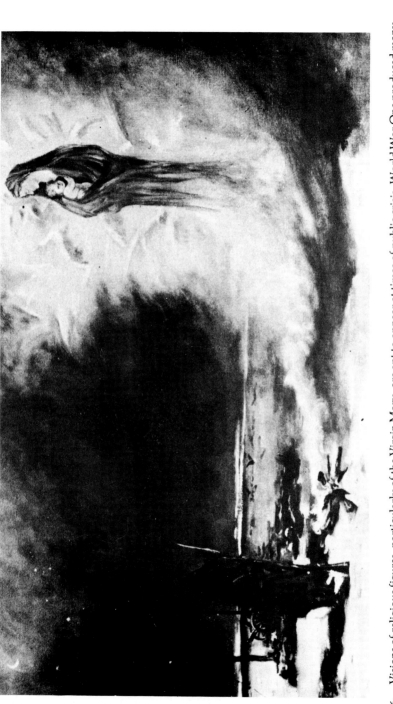

6. Visions of religious figures, particularly of the Virgin Mary, are apt to appear at times of public crisis. World War One produced many such reports. This Madonna entity was seen on the Russian front in 1915 (see page 116). From a painting by the Revd B. S. Lombard.

7a. Identification of visionary entities is not always straightforward. The figure seen by Bernadette Soubirous (see page 106) did not make its identity clear, and the eventual identification was circumstantial and ambiguous. Certainly it resembled a young girl rather than the traditional figure depicted in this pious engraving from a souvenir book of 1899.

7b. The entity seen at Fatima in 1917, as described by the witnesses and depicted here, was very different both from the traditional Mary-figure and from the representations now on display at Fatima. From Jaoquim Fernandes and Fina D'Armada, *Intervençao extraterrestre em Fátima.*

8a. Though if he chose he could disguise himself any way he wished, the Devil traditionally appeared in archetypal form, as in this anonymous engraving illustrating a French folk-tale.

8b. Early woodcuts, too, show the Devil neglecting the tricks at his disposal and appearing to his victims in unmistakeable guise. Perhaps he knew that his admirers would admire him none the less — perhaps all the more — in his natural form. From Sprenger, *Malleus Maleficarum.*

9. A modern 'Man in Black' as drawn by a man who claimed to have
encountered three of them. This is Albert Bender's drawing of one of the trio
who visited him in August 1953. (See page 139.) From his book *Flying Saucers
and the Three Men*.

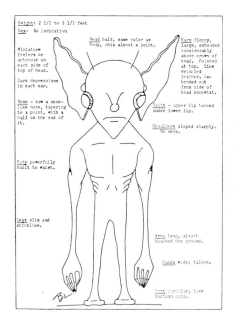

Height: 2 1/2 to 3 1/2 feet
Sex: No indication

Head bald, same color as Body, chin almost a point.

Miniature feelers or antennae on each side of top of head.

Dark depressions in each ear.

Nose - saw a cone-like nose, tapering to a point, with a ball on the end of it.

Body powerfully built to waist.

Legs slim and sticklike.

Ears floppy, large, extended considerably above crown of head. Pointed at top. Like wrinkled leather. Extended out from side of head somewhat.

Mouth - upper lip tucked under lower lip.

Shoulders sloped sharply. No neck.

Arms long, almost touched the ground.

Hands wide; talons.

Feet circular, like suction cups.

10a. UFO-related entities are popularly supposed to be 'little green men'. One of the rare instances when they matched this description was at Kelly, in Kentucky, on 21 August 1955, when the Sutton family believed themselves attacked by a number of entities. This drawing by Bud Ledwith was based on a witness description at the time.

10b. Most contactee stories involve encounters with human-like figures, who tend to be somewhat androgynous in appearance and are invariably kindly, loving and caring. This drawing is from *The book of Spaceships and their relationship with Earth,* allegedly written by 'the God of a Planet near the Earth' about 1960.

11. Folklore entities range from 'little people' of many kinds, usually easily recognizable, to naturalistic figures easily mistaken as human. The Norwegian *hulder*, a local variant on the universal other-worldly seductress entity, relied on being mistaken by her victims for an ordinary country girl. Illustration by unnamed nineteenth-century artist for Asbjörnsen's *Huldre-eventyr*.

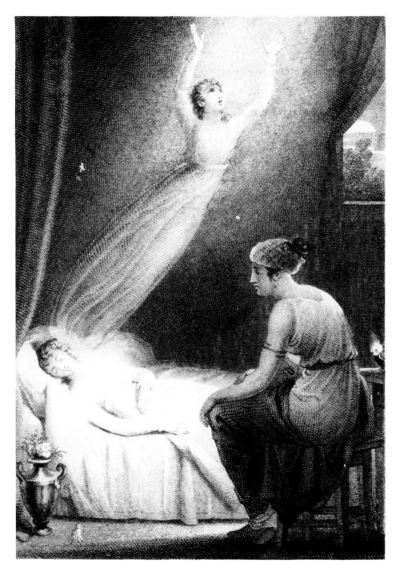

12. There is widespread belief that we each of us possess an 'astral body' that leaves our physical body at death — and also occasionally during life. It is generally supposed to be an exact facsimile of the physical body, as in this steel engraving by Corbould, done in 1829.

13. Conjuring up spirits is a traditional test of a magician's capability. Here the renowned sixteenth-century necromancer Edward Kelly, with his companion Waring, successfully disturb the rest of a spirit in the churchyard at Walton-le-Dale in Lancashire. From an aquatint in *The Astrologer* (1826).

14a. Remarkable spirit manifestations were reported as occurring nightly at the Eddy homestead in Vermont in 1874 (see page 226). One of those sent to look into the reports was Henry Olcott: this drawing by Alfred Kappes, made on the spot for Olcott's subsequent report *Report from the other world*, published the following year, shows a typically dramatic confrontation between the living and the dead. It is captioned 'The reunited family'.

14b. Human gullibility has not diminished with the increasing sophistication of modern life. This photograph shows a figure identified as the returned spirit of Queen Astrid of the Belgians, who had been killed in a motor accident on 29 August 1935. The alleged materialization took place during a seance at København on 31 May 1938, through the mediumship of Einer Nielsen. From Gerloff, *Die Phantome von Kopenhagen*.

15. 'Bedroom invaders' can take many forms, though they are predominantly female. The most notorious example is the succubus, depicted here by Robin Ray.

16. Pictures of percipients in the act of having an entity-sighting experience are understandably rare. However, the visionaries of Garabandal in 1961 were frequently photographed. This photograph shows Maria-Dolores, Conchita, Jacinta and Maria-Cruz in July 1961, in ecstasy before a vision of the Virgin. The simultaneity of their actions, regarded by some as an indication of the miraculous character of the event, is however characteristic of a known, if infrequent, psychological process known as 'folie à deux'. From F. Sanchez-Ventura y Pascual, *Las apariciones no son un mito* (1966).

For the purpose of our inquiry, it is not crucially important that there should have been an original real-life case: what is important is that the legend existed, that there was an archetype that one or more of the four travellers may have known of or subconsciously remembered. For however we account for the Montpellier case, there can be no question that it relates to the pattern of all the other phantom hitchhiker cases that have been reported over the years, whether or not they had any basis in fact.

The phantom hitchhiker, more than any other type of entity-sighting, brings us up against the paradoxical nature of reality. For *any* explanation we devise for what happened at Montpellier in 1981 involves a process that defies all our notions of cause and effect. Not only did something 'impossible' occur, but it occurred in a way which turns even logic upside down.

★ ★ ★

Every culture has its own stock of traditional lore, in which a range of more or less stereotyped entities tend to recur. The 'man in black' and the 'women in white' are two types which appear in one form or another in every country at every period: the men in black tend to be associated with forces of evil — devils and demons — while the women in white are often wronged women or benevolent spirits. These are not universal stereotypes, however: there are malevolent females too, some of which appear as hags, who look as evil as they are, others who appear benevolent but are deceptive, such as the *hulder* of Norway, one of countless forms of the *femme fatale* who lure simple-minded males to their doom.

In his classic study *The Fairy Faith in Celtic Countries*, [165] Evans Wentz considered the various explanations available at the time (1911) for the origin of the belief in fairies. There was the *naturalistic theory*, which held that the belief was the result of primitive man's attempts to explain or rationalize natural phenomena — fairies incarnate natural forces and are responsible for the vagaries of the environment. There were two theories that proposed that the beliefs were simply *folk-memories*, of an actual but now vanished pygmy race, some suggested, or of the druids and their magical practices, as was believed by others. Fourthly there was the *mythological theory*, which saw the fairies as diminished versions of the old pagan divinities worshipped by Celtic ancestors.

Elements in some of these models Wentz found acceptable so far as they went, but none of them seemed to him adequate as a total explanation. As he pointed out, it is all very well to say that today's fairies are yesterday's nature-goddesses, but this still leaves an explanation to be found for how the nature-goddesses themselves came into being. So he was led to formulate a root-theory, which he named, loosely and ambiguously, the *psychological theory* of fairy origins. It embodies the following principal propositions:

1   Fairyland exists as a supernormal state of consciousness into which men and women may enter temporarily in dreams, trances, or in various ecstatic conditions.

2   Fairies exist, because in all essentials they appear to be the same as the intelligent forces now recognized by psychical researchers, be they thus collective units of consciousness like what William James has called 'soul-stuff', or more individual units, like veridical apparitions.

Writing in 1911, Wentz can be forgiven for the uncomfortably loose formulation of his theory; but we, with seventy years of study to draw on, cannot permit ourselves such vagueness. On the other hand, can we say with any degree of confidence which of the models currently available Wentz would recognize as being the closest match to what he had in mind? However, if Wentz's theorizing is to be of any help to us, we have to reformulate it in more sophisticated terms. At the risk, then, of discerning meanings in his text which are not really there, I suggest that an updated version of Wentz's belief would be roughly as follows:

> Fairies are real entities, with at least a quasi-physical nature and possessing the capability of autonomous action; they are, however, dependent for their capability of being perceived — and perhaps for the physicality of their existence altogether — on some subconscious mental process on the part of the percipient. They are therefore likely to be seen when the percipient is in a special state of consciousness, such as dream, trance, or ecstasy. In such states, the percipient gains access to a parallel plane of reality, where these entities are imagined as having their existence.

When expressed in these terms, the parallels with others of our entity-seeing categories become obvious. The concept of fairyland, for example, is the result of providing the fairies with a 'home', because we humans are conditioned to the belief that 'there is a place for everything': on the same principle, the Virgin Mary visits us from her heavenly home, and our Cosmic Saviours originate on some distant planet that we could in principle detect with our telescopes. Similarly, the interdependence of fairies with humans, which is so marked a feature of all fairy beliefs, may be seen as an unconscious recognition of the fact that fairies are, indeed, totally dependent on human thought-processes, so that Barrie's Peter Pan was literally correct in saying that 'every time a child says "I don't believe in fairies" there is a little fairy somewhere that falls down dead'.

I am very conscious that I may have done no more than rewrite Wentz's ideas in terms of current paradigms, to then exclaim with astonished delight how well his thinking fits into our more advanced formulations.

I do not think I have twisted his intentions, but in case I have unwittingly done so, I shall regard his contribution as provisional only. However, I think there can be no doubt that, whatever terminology he might have adopted, the general thrust of Wentz's ideas was along very similar lines to those along which our own hypothesis of entity-origins is developing.

It is significant, too, that Wentz himself was constantly aware of links between his field of study and other types of entity. He specifies 'veridical apparitions' on the one hand, and refers to the investigations of psychical researchers on the other. At the same time he was aware of the borrowings from the fairy-faith that had found their way into Christian belief, and it would have been obvious to him that the entity seen by Bernadette at Lourdes was a cousin, and not a very distant one, to the fairy-entities that his witnesses were seeing in their Celtic homelands.

One of the scenarios Wentz had to consider was that these entities may be the outward forms of beings that are not fairies, not hitchhiking ghosts, not alien visitors, not the Virgin Mary, not Aunt Agnes, but some kind of elemental being that takes on these guises for purposes of its own.

There is nothing new about this hypothesis, which is prevalent throughout folklore and has received a new lease of life with the rise of science fiction and the concept of parallel universes, which become a convenient place of origin for such beings. Religious interpreters of the UFO phenomenon, in particular, are much inclined to look for explanations along these lines. The idea would be that the discarnate force — which may be conceived as of benevolent, malevolent, or simply neutral — takes on the specific form of an entity that has meaning for the percipient, with the aim of beguiling, seducing, encouraging, warning, informing, frightening, arousing remorse or guilt, or — why not? — simply of amusing him.

The theosophist Edward L Gardner, who firmly believed in the physical existence of fairies, gave this account of their nature, which is sufficiently intelligent for us to give it serious consideration despite the temptation many of us must feel to dismiss such an account as being hopelessly fanciful:

> It must be clearly understood that all that *can* be photographed must of necessity be physical. Nothing of a subtler order could in the nature of things affect the sensitive plate. So-called spirit photographs, for instance, imply necessarily a certain degree of materialization before the 'form' could come within the range even of the most sensitive of films. But well within our physical octave there are degrees of density that elude ordinary vision. Just as there are many stars in the heavens recorded by the camera that no human eye has ever seen distinctly, so there is a vast array of living creatures whose bodies are of that rare tenuity and subtlety from our point of view that they lie beyond the range of our normal senses.
>
> Fairies use bodies of a density that we should describe, in non-technical

language, as of a lighter than gaseous nature, but we should be entirely wrong if we thought them in consequence unsubstantial. In their own way they are as real as we are.

The normal working body of the gnome and fairy is not of human nor of any other definite form. They have no clean-cut shape normally, and one can only describe them as small, hazy, and somewhat luminous clouds of colour with a brighter spark-like nucleus. Instantly responsive to stimulus, they appear to be influenced from two directions — the physical outer conditions prevailing, and an inner intelligent urge. The diminutive human form — sometimes grotesque, as in the case of the brownie and gnome, sometimes beautifully graceful, as in the surface-fairy variety — if conditions allow, is assumed in a flash. For a while it is retained: if disturbed or alarmed the change back to the slightly subtler vehicle, the magnetic cloud, is as sudden as the birth. What determines the shape assumed and how the transformation is effected is not clear. One may speculate as to the influence of human thought, individual or in the mass, and quite probably the explanation when found will include this influence as a factor — but I am intent here not on theorizing, but on a narrative of observed happenings. [49]

Besides his own 'observed happenings', which appear to have been widespread, those of several other witnesses are quoted in Conan Doyle's book, in which the Gardner hypothesis is set forth. The following case was supplied by the well-known author (well-known, moreover, as a debunker of legends) Sabine Baring Gould:

When my wife was a girl of fifteen, she was walking down a lane in Yorkshire, between green hedges, when she saw seated in one of the privet hedges a little green man, perfectly well made, who looked at her with his beady black eyes. He was about a foot or fifteen inches high [30-40cm].

Another first-hand report came from the author Violet Tweedale:

One summer afternoon I was walking along the avenue of Lupton House, Devonshire. My eye was attracted by the violent movements of a single long blade-like leaf of a wild iris. Expecting to see a field-mouse astride it, I stepped very softly up to it. What was my delight to see a tiny green man, about five inches long, swinging back-downwards. His tiny green feet which appeared to be green-booted were crossed over the leaf. I had a vision of a merry little face and something red in the form of a cap on the head. For a full minute he remained in view, then he vanished.

These are not isolated instances: there are many more in the literature, including several of quite recent date. It would certainly seem as though a good many witnesses sincerely believe they have seen this kind of entity, and there is at least one case on record in which two witnesses saw the same entity at the same time. Once again, we have to face the paradox that real life is imitating art, for it is notable that all these sightings are of a traditional 'fairy story' type of creature.

A hypothesis not unlike Gardner's was put forward — very tentatively — by researcher Maxwell Cade, when considering the UFO-entity phenomenon. He speculated:

> Suppose that they have equipment (mechanical or biological) which can track our thoughts? The landing vehicle receives from a nearby group of humans some confused thoughts and emotions ('A flying saucer . . . FEAR . . . perhaps they are giants, or hairy monsters . . .'). No sooner received than the reflected thought-image is on its way. The human observers are met by just the kind of alien creature they feared to meet.[15]

Cade goes on to draw a useful parallel with the materializations of the seance room, where so often the observed — or seemingly observed — phenomena are precisely what the sitters expected to see. His hypothesis has the great merit of reconciling two features of such sightings — the great diversity of the UFO-related entities, and their conformity to the limits of human imagination: and of course the same is true of Gardner's fairy hypothesis. Moreover, there is some support for Cade's seemingly far-fetched suggestion: recent UFO investigation has confirmed, apparently beyond doubt, that some types of UFO are able to pick up the thoughts of observers, if only to a very rudimentary degree.[40]

Such suggestions also account very neatly for cultural variations in entity-sightings. The traditional British fairy, of the gossamer wings and dainty skipping kind, is a far cry from the fairies of Scandinavia, which are much less decorative and generally nastier. Presumably Gardner would argue that each culture sees what it expects to see, and in this respect entity-sightings represent a cultural feedback. The 'nice' British fairies and the sinister Norwegian variety would be the expression of differing impulses at work in the respective collective unconsciousnesses of the two races.

Despite the attractions of these hypotheses, it must be admitted that they are wildly speculative. The notion of an amorphous entity, which assumes a specific shape in response to human thought, is one that can usefully be applied to many of our entity categories: but it seems to be a far cry from the hallucination and the hypnagogic image. For the moment, we must simply set it on one side as just one more possible mechanism.

Within the last few years, however, a new biological model has been proposed that could not only provide a solution to our dilemma, but do so in acceptably scientific terms. This is the process of *morphic resonance* introduced by Dr Rupert Sheldrake.[141] His hypothesis, as I understand it, is that the form, development and behaviour of a biological organism are largely determined, not only by its heredity and its environment, but also by a morphogenetic 'field', which allows it to 'communicate' with other organisms of its nature. As a result, the experience of one is shared, or at any rate known, by another, and this permits common processes of development and adaptation even when one organism is far separated from the other; furthermore, the separation can be in time as well as space.

Sheldrake's hypothesis of formative causation, applied to our material, could enable us to construct a model whereby different percipients, widely separated in space and time and with no communication along conventional channels, might have entity-sighting experiences that had far more in common than coincidence would warrant, yet hardly seem traceable to any common origin. Could the phantom hitchhiker, for example, be an instance of formative causation operating in a purely behavioural context?

Since we have not arrived at even a provisional hypothesis as to what agency is causing such events to occur, it might seem premature to be considering the mechanics of the process — except that if we knew *how* the thing was done, it might help us to know who is doing it.

What, in practical terms, are we suggesting? Very loosely, that the agent responsible for causing these events is aware of the 'idea' of such an event — which could be, for example, a real-life abduction by extraterrestrials, a real-life warning uttered by the ghost of an accident victim, a real-life visitation from an entity which is, or is impersonating, the Virgin Mary, a real-life battlefield vision of helpful angels; it could also be an imagined fantasy of any of these things, or even a consciously fabricated account of them. It really does not matter, so long as the *idea* is somewhere in existence — in print, in oral tradition, in the conscious mind of an individual or even in his unconscious. Its existence in *any* form is sufficient for it to be capable of being detected, and *used*, by some other agency.

This may of course be a totally illegitimate extension of Sheldrake's hypothesis, and in any case his ideas are highly controversial and likely to remain so until extensive testing has validated/invalidated them. Nevertheless, they offer us a new way of looking at our material, and suggest ways of resolving one of the most puzzling aspects of the material: the disconcerting degree of similarity between reports made by percipients who could hardly have had knowledge of one another's experiences, and even more, those cases in which real life seems to be modelling itself on an abstract idea.

1.13 SPONTANEOUS CASES: A PRELIMINARY ANALYSIS

We have now looked briefly at the various kinds of entities people claim to have seen spontaneously — that is, that came unsought and generally unexpected. In the next part of our study we shall be looking at entities seen as the result of deliberate efforts on the percipient's part, with the intention of establishing whether or not they constitute a wholly separate class of phenomena. It is too soon to make a definitive analysis of what we have established hitherto, but a preliminary analysis is justified, first, to get it clear in our minds just what the problems are, and second, to enable us to ask the right questions when considering the experimental phenomena of Part Two.

I think it is clear that we have established that we are speaking of a wide range of phenomena. Some critics would say that I have cast my net too wide: but I think the evidence does not permit us to set aside totally any of the categories we have considered. While it is true that each of the categories has its own characteristics — which is why we are able to consider it as a category at all — it is equally true that there are many common features, shared by more than one and sometimes by most of the categories. I have done my best to maintain the traditional distinctions, and have sought to establish definitions of individual categories that indicate their specific characters with as much precision as possible. Yet it must be as obvious to the reader as it is to myself that absolute criteria are impossible to establish. Into which category, for example, would you put the case that follows?

A British climber, while climbing at 6000 metres in the Himalayas, suddenly came upon two friends from his schooldays, both of whom had been killed in a motor accident a dozen years earlier. [42j]

Griffith Pugh, the doctor who reports the case and himself a mountaineer, takes the view that such sightings are 'only hallucinations caused by extreme cold, exhaustion and oxygen shortage even when breathing equipment is being used'. No doubt he is right, but his phrase 'only hallucinations' begs a very large question. For as we have seen, the majority of hallucinations are of unidentified persons, whereas these were apparitions of recognized dead persons. Appearing as they did, the entities were so manifestly 'out of place' that it is unlikely the percipient even for a moment thought they were really there. Had the sighting occurred in different circumstances, the percipient might well have felt himself to have been visited for some reason related to the apparents. Even as it is, the possibility exists that the apparents came of their own volition, to give their old friend comfort and encouragement; in which case his cerebral anoxia did not *cause* the experience so much as enable it to occur.

Such a case raises the question that underlies our entire inquiry: who, in this kind of experience, is responsible for the sighting — the percipient, the apparent, or some external agent? Or is it a shared operation to which both percipient and agent contribute?

Here is another borderline case:

In a household in Ohio, USA, several young daughters shared a bedroom. One night ten-year-old Mary went in to her mother, who slept in a neighbouring room, complaining that her older sister Nancy had come and stood by her bed and wouldn't go away. The mother went back with Mary, saw that Nancy was asleep in her own bed, and concluded that Mary had imagined the whole thing. But before long Mary was back again, complaining that this time Nancy was again standing by her bed, but would not answer when spoken to. Once again the mother found Nancy asleep. Throughout the night, at intervals, Mary kept talking to Nancy, asking her to please go back to her own bed.

The next morning everyone got up but, as was habitual, they left Nancy sleeping, for she was not a strong child and needed extra sleep. Later, at bed-making time, her mother called her; getting no response, she came closer and found that she was dead. The doctor's opinion was that she must have died soon after going to bed.[127]

The ostensible explanation — that Nancy's 'ghost' was trying to communicate with Mary — is only one way of accounting for this little incident whose very simplicity conceals any number of puzzles. Did Nancy (whether she was dead or alive) know what she was doing? Why was she doing it? What part of her, if any, was standing there by Mary's bed? Why did Mary alone see her? Was Mary's subconscious playing any part in the manifestation? Why couldn't Nancy respond to Mary?

With so much unanswered, the question whether Mary was 'hallucinating' or seeing a 'ghost' or an 'astral double' becomes mere playing with words that are as likely to confuse the issue as to illuminate it.

Is there anything we can usefully say about the whole spectrum of phenomena considered as a body of experience?

First, we must not lose sight of the one thing they all have in common, that all are perceived visually: that is, the percipient has the clear impression of having a visual experience in which his visual senses are involved. This is no less true even if he is simultaneously aware that he is not having a 'normal' visual experience — that what he is seeing is somehow unreal, whether because of what it is (such as a person he knows to be dead) or because of the way it appears (suddenly manifesting through a wall). Though he may not believe his eyes, he does not usually question that his eyes are involved.

Second, this subjective impression is matched by the objective evaluation of the experience, which is forced to conclude in every case that it is in one or another sense of the word illusory; that is, the thing seen is not real in the way we usually use the word.

It is this conflict between subjective impression and objective evaluation that is the central issue in all these types of experience, and the element that justifies us in ranging them side by side. However different in kind the experiences, the process of having the experiences may not be so different. And when we turn to the nature of the sighting experience itself, we find that there are aspects that can be analyzed to enable us to formulate specific lines of inquiry:

★ The apparent can be of three kinds:

1   *Known persons.* These are perceived in most dreaming cases, doppelgängers, some apparitions and most hauntings. Sometimes they are close friends or relatives, sometimes they are comparatively distant acquaintances or even public figures such as the Queen: sometimes they seem to have significance for the percipient, sometimes not.

2   *Stereotypes.* These are always seen in visionary and demonic sightings, and in folklore cases: it seems likely that Men in Black and UFO-related cases also involve stereotypes rather than specific alien visitors. Some apparitions give the impression of being stereotypes also.

3   *Unknown.* The great majority of hypnagogic and hallucination cases seem to be of strangers, and a majority of companions and counsellors. It is virtually impossible to establish that the entities are truly strangers — they could be people sat next to once in a bus or seen long ago in a crowded store, not consciously recalled but nonetheless stored in the memory.

★ The *circumstances* of the sighting range from total unawareness to total awareness. Dreams, by definition, are seen only when asleep (I exclude so-called 'waking dreams' because I do not regard them as the same phenomenon and see the term as a confusing misnomer). Hypnagogic visions, again, occur by definition only when on the verge of entering or leaving the sleep state.

A good many experiences occur when the percipient is in a crisis situation — when danger, confusion, emotional factors and the like may be supposed to have put him into a special state which in turn may be supposed to be somehow conducive to seeing an apparition. The question remains whether the crisis *puts the percipient into a state where he is receptive* to manifestations which in the normal state he would not have perceived; or whether his critical state *causes* the manifestation.

Many percipients speak of being in a relaxed, drowsy state when having their experience, and again we may suppose that this kind of state creates a receptivity that is not normally operating. But this is very far from being a general rule: at least as many events occur when the percipient is, so far as can be ascertained, in his normal alert state.

★ Evidence of *purpose* is clear in some instances, totally absent in others. Dreams, doppelgängers and apparitions may bring warnings, and so in their own way do Men in Black and UFO-related entities. Visionary entities bring messages to humanity at large as well as personal advice and exhortation: sometimes their requests seem incompatible with their alleged identity (for example, the desire of the Virgin to have prayers said and chapels erected in her honour seems 'out of character'). Some apparitions and a good number of hauntings appear to have purposes connected with the apparent, something unfinished during the earthly life of someone now dead, for example: companions, on the other hand, however autonomous, seem to be there for the unique purpose of comforting and counselling the percipient.

In short, we may say that, under a variety of circumstances, percipients have the illusion of seeing entities which, either at the time or subsequently, they know are not 'real' in the accepted sense, even though they may be perfect replicas of real persons. These sightings often occur at times of need on the percipient's part, suggesting that it is he who initiates the sighting; but also they often occur at the time of crisis on the apparent's part, suggesting that it is *he* who is responsible. So it seems we have to hypothesize a process which can be initiated either internally or externally.

In Part Two of our inquiry we turn to cases in which the sighting is not simply initiated by the percipient, but is *deliberately* initiated.

PART TWO

EXPERIMENTAL ENTITIES

INTRODUCTION

In the first part of this study we briefly surveyed a number of different kinds of experience in which people reported seeing entities of a more or less human appearance. The *circumstances* in which these experiences occurred would alone suffice to indicate that we are faced with a variety of phenomena: over and above this, we have found that there are inherent differences in the *nature* of the experiences themselves, which are often specific to the type of experience. For example, the nature of the imagery seen in hypnagogic states seems to be unique in several respects, so that it is reasonable to conclude that people can only have that kind of experience if they are in that state.

Moreover, there are kinds of experience that are unique in their *subject;* there are visions of religious figures, such as the Virgin Mary, that seem to appear only to limited sections of the populace, and apparitions manifesting to friends and relatives where the state of the percipient, is clearly less important than the circumstances of the apparent.

But we have also noted parallels: however different they may be in other respects, these experiences have some elements in common. Though the end results may be distinctive, the processes involved may be much the same. It seems justifiable, therefore, to continue to treat the experiences, as far as possible, as variations on a single theme as we proceed with our inquiry.

For there is still much more evidence to be weighed. This evidence comes from experimental procedures which have been deliberately set up, controlled observations of phenomena, and practices undertaken deliberately in the hope of replicating events that have occurred involuntarily, such as spiritualist spirit-evocation and magical spirit-conjuring. Other evidence consists of parallel experiences which, while not specifically related to sightings of entities, do seem to suggest mechanisms or processes that could also be operating in entity-sightings.

Some of this parallel evidence is supportive of the view that entity phenomena have some kind or degree of autonomous existence; other findings seem to indicate the contrary. But between them they add substantially to the quantity and quality of data available, and will put us in a much better-informed position when, in Part Three, we start to consider explanations and models.

2.1 IMAGINARY CONTACTS

In 1971 Brian Scott, a 28-year-old American, underwent an experience that became known as 'the Garden Grove case'. Bewildering as it was to Brian himself, the case was not particularly different from many in which seemingly sincere persons have reported encounters with alien entities who seem to come from outside our Earth. As is usual in such cases, there were some who were ready to accept Brian's account at face value, others who were inclined to be sceptical and suspect a hoax, and yet others who respected the man's sincerity but found it easier to believe that some kind of psychological process had been involved. There were, in addition, some who believed that 'powers of darkness' were responsible.

Fortunately, one of those investigating the case was John DeHerrera, who was working on behalf of APRO, one of the more responsible American UFO investigative organizations. He saw that the Scott case presented certain features that could be tested scientifically. With the assistance of Alvin H. Lawson, a professor of English at California State University and another resident of Garden Grove, and Dr W.C. McCall of Anaheim, California, DeHerrera initiated an experiment that was to have profound implications for the whole subject of alleged contact cases, and which is relevant to an even wider range of reports. It is not too much to say that the 'imaginary abduction' experiments require all of us to rethink our notions of the nature of reported experience. Not that they necessarily invalidate earlier thinking, but they show that the whole matter is even more complicated than has been hitherto supposed.

The original incident, briefly, was as follows:

In March 1971 Brian Scott, with a companion Eric Wilson (pseudonym), were camping in the Arizona desert, on a trip to catch snakes on behalf of a local university researcher. Sitting talking outside their tent one evening around 9 p.m., they heard a coyote in the distance; looking in the direction of the sound, Brian saw a brilliant object hovering near the mountains, approximately eight kilometres distant. The object was oval, dark, but surrounded by a bright haze, and with a bright purple or purplish-green light beneath. It was silent, and did not seem to be a conventional aeroplane or a helicopter. Brian pointed his torch at the object, despite the distance: this seemed to cause it to move towards them, for the light seemed to grow larger. When it was almost directly overhead it was so large that it blocked out the sky almost entirely. It hovered over them at a height of about sixty metres, bathing them in its purple light. They could see no windows or doors, nor any other features of a conventional aircraft. Naturally, the young men were frightened: 'I wanted to run but something was holding me,' Brian later reported.

He remembered nothing more until he was driving home: he had no recollection of any intervening events. An announcement on his car radio told him it was 11 p.m., so that he seemed to have spent two hours that he could not recall. The decision to leave their camp and expedition seems to have been made unconsciously.

Some four years later he told a fellow-worker of his UFO experience: his colleague advised him to tell UFO investigators of it. He did so, and this led to his being put under hypnosis in 1975, in the hope that in this state he might be able to retrieve more details of what had happened to him.

Despite some character defects (Brian was a noted practical joker with a casual attitude to life, and, more seriously, had once served a six-month prison term for taking money for work he had never carried out), it was felt that his case was suffiently interesting to justify in-depth investigation. The effort seemed to be justified when the session began to show results. Under hypnosis, Brian now revealed that when the UFO had been over their heads, he and Eric had been lifted off the ground towards a hitherto unnoticed door in the UFO. They were taken through this door into the craft, where they encountered figures about half a metre taller than himself, with ugly, wrinkled skin and very large arms ending in three-fingered hands, and fat, short fingers. He could not describe their feet so precisely because there was a kind of fog at floor level.

The two young men were undressed and taken away: Brian said he felt as though he was being carried. He was taken into a very brightly lit room. He tried to struggle, but ineffectively. He then had an experience that he found hard to describe, as though his body was being taken apart; evidently it was some kind of physical examination. Then one of the entities, about three metres tall, who gave the impression of being the leader, came over to Brian, put his hand on his head, and conveyed, seemingly by telepathy, the message 'I am of another world — beyond all time, through years of light years.' Brian felt a sense of reassurance, that he need not be afraid. A feeling of calm and happiness was induced; he had a vision of mountains, and then of a street peopled with calm, happy figures. 'All men will know the truth,' the message continued, 'None is to fear. The world is not known to all. No further away than time. When I return, all will know the truth. Life will be lived through all time and knowledge will all be yours. I shall return to you and all mankind. You will see your world, my world, and the joy of all life to come . . . not measured by time of man, only as light reflects light, time reflects time, and a mirror reflects what you see now. To cell from body. To you. I am now, I was, and always will be.'

Subsequent hypnotic sessions added little further detail, but included the prediction of a colossal catastrophe involving the 'complete annihilation of the Western hemisphere' from an 'easterly direction', involving 'bombs of high magnitude'. From this disaster only a few thousand persons would be rescued and taken to another plane of existence. All this is due to occur on 24 December 2011.

Clearly it would have been helpful if Eric had been able to provide an account to corroborate his companion's. Unfortunately he was reluctant to become involved, refusing to admit that he had even so much as seen the UFO, let alone that any abduction had taken place. When asked, he thought it possible that his friend had made the whole thing up.

Eventually, he was persuaded to subject himself to hypnosis. In this state, however, Eric continued to insist that he had seen no UFO. Then the hypnotist suggested to him that, setting aside what had actually happened, he should use his imagination and describe what *might* have happened. 'I want you to use your imagination now. I want you to pretend like you can see these things.' Somewhat diffidently, the hypnotized Eric began to fabricate a story. But though he now agreed that there had been a UFO, and that he had been frightened by it, that was all.

The question still remained unanswered: had Brian had a genuine experience? While in the hypnotic state, he was repeatedly urged to produce writings and drawings to supplement his verbal account, and eventually Brian obliged: they turned out to be largely incoherent and seemingly meaningless. The implication seemed to be that the young man's subconscious was creating them simply to go along with the investigators' request; but, having no authentic material from which to fabricate the material, Brian was doing his inadequate best to create it from nothing. About the same time, Brian tattooed himself on the forearm with two figures, a spider and a jaguar, which he claimed he had designed according to instructions received while he was in a trance state. Curiously, they are very similar to the huge, mysterious designs made on the ground at Nazca, in Peru, whose significance remains obscure but which are often cited in connection with UFOs.

From this stage on, the Scott case grew more and more confusing. Further abductions were reported in 1973 and 1975. A variety of poltergeist-type phenomena were reported, notably balls of light which appeared both indoors and in the garden of his residence, and were seen both by the family and by visitors. Other such lights were reported in the neighbourhood. Strange sounds and smells, spontaneous combustion of bed-covers, the sighting of a vaporous, human-like figure on several occasions, were among the phenomena reported. In addition, Brian's own behaviour became erratic, which was attributed to the tension created by these events. He would disappear from home for twenty-four hours, returning home filthy: he sought publicity from various groups of a cult nature. In 1977, encouraged by others, he laid plans for a movement named the Congress for Inter-planetary Technology and Education, purportedly basing it on New Age information revealed to him by his abductors.

At the same time, in continuing hypnotic sessions, Brian would frequently go into a state in which he seemed to be in communication with some kind of demon. On occasions he would seem himself to be possessed, identifying himself as Beelzebub: in this persona he on one occasion offered his sister-in-law power, riches and good health in return for her soul. Another characteristic of his 'revelations' was continued reference to South America, where he insisted that further developments could be expected.

In many of these manifestations he was suspected by the investigators of faking, consciously or otherwise: for example, when he came home from a park with a carved spider which he had 'found' and which he claimed to be a sign from his abductors, it seems likely to have been a deliberate fabrication. To anyone at all versed in the accounts of such cases, what was happening to Brian, bizarre as it was, was by no means unique: there are literally dozens of cases equally complex and equally ambiguous, sharing such noteworthy features as:

★ The isolated situation of the witness when the initial experience occurs.

★ The conscious awareness of a UFO experience but no memory of an abduction.

★ The recollection under hypnosis of a UFO abduction, which itself runs close to a pattern in which the same details crop up time and again.

★ The insertion of material from other fields of the occult, such as demons, South American antiquities and mysterious 'signs'.

★ The occurrence of seemingly genuine psychic phenomena, such as the poltergeist-type balls of light, which seem to have been entirely genuine and testify to a real emotional crisis, no matter how caused.

★ The claim to have been specifically contacted by alien entities, to have been entrusted with some kind of mission, and the consequent formulation of a 'philosophy' embodied in a cult, group or some such organization.

★ Psychological effects in the witness which, unfortunately, cannot necessarily be identified as consequences of his experience because, no psychological evaluation of him having been made prior to the sighting, there is no certainty that they were not in him, latent, all the time.

★ Total absence of any material proof, even in the form of testable information.

Among the many later events was an occasion on which Brian returned to the desert site of his first alleged encounter: on his return he claimed to have been abducted a second time. This was easily revealed as a hoax by investigator DeHerrera. The incident confirmed the baffling ambiguity of the case in which sincerity and falsification seemed inextricably mixed. How could the two be reconciled?

The answer came not from Brian's testimony but from his friend Eric's. DeHerrera asked himself what significance there might be in the fact that Eric provided an account of the UFO sighting — admittedly, only a very sketchy one — only when invited to imagine a fictitious event. It was this which suggested an experiment which proved to be highly significant, even if its implications have yet to be fully appreciated.[30]

By means of an advertisement in a college newspaper, Lawson and McCall sought volunteers for an unspecified hypnosis experiment: the advertisement requested 'creative, verbal types'. Several volunteers presented themselves: these were screened down to an eventual eight who had — so they claimed — read no UFO literature and knew virtually nothing about the subject. Clearly, these days, there is hardly anybody who is totally unaware of the UFO phenomenon: but at least there are some who are only minimally informed.

The volunteers were kept separate during the tests, and the experiment started by their being asked to imagine themselves at some location — a beach, mountains, desert, etc. — which they enjoyed visiting. Under hypnosis, they were asked to visualize themselves at this chosen location. They were then asked eight questions in succession:

1   *There is a UFO travelling towards you: would you describe it?*

2   *Imagine you are aboard the UFO. How did you get aboard?*

3   *Describe what you see inside the UFO.*

4   *Imagine you see some entities or beings on board the UFO. Describe them as completely as you can.*

5   *You are undergoing some kind of physical examination. Describe what is happening to you.*

6   *You are given some kind of message by the occupants of the UFO. Describe that message and how it came to you.*

7   *You are returned safely to where you were before. How did you get there and how do you feel?*

8   *Imagine it has been some time since you had that encounter. Were you affected physiologically and/or psychologically?*

These questions were based on the allegedly true accounts which have been furnished by people who claim to have been subjected to genuine abduction incidents. While they are admittedly 'leading' questions, in the sense that they dictate the general direction of the subject's response, they do not propose any specific scenario and leave room for a wide range of interpretation. It should be remembered that the object of the experiment was to ascertain not *what* happened, but *how* it happened.

The results surpassed the expectations of the investigators. The imaginary encounters were richly packed with detail, as opposed to vague generalities: and furthermore, that detail conformed closely with the detail provided by the allegedly 'real' abductees. The 'real' and the imaginary abduction experiences were wellnigh indistinguishable. Examples of the kind of detail provided by the volunteers are:

'A long tube came out of [the UFO] . . . it was about two feet above me . . . this long cylinder-like tube came down. It was grey and . . . was like all coloured lights inside of it . . . I seemed to be floating for a second, and then I was inside.'

'I only see a head, and it's very strange, and then the more I look at it, the more it actually becomes more human as I look at it . . . It begins to conform to what I want to see as being human. I wonder if it's giving me this illusion of itself, strictly through my mind . . . I get the feeling that it's changing for me . . . I think I'm seeing a projection of one of them . . .'

'They're using light — pinhole beam of electricity that doesn't shock . . .'

'There's no verbal communication, just thought . . .'

These brief excerpts indicate that the accounts were rendered with a considerable degree of sophistication: given the bare questions quoted above, there was no requirement or inducement for the subjects to elaborate to this extent. So one of the first questions must be: what was it that encouraged the subjects to indulge themselves in such highly developed fantasizing?

What struck the investigators most forcibly was that the details frequently conformed to those they had previously noted in the 'real' accounts, which had not been communicated to the 'imaginary' subjects, of course, but which had been noted as a kind of checklist of details that might crop up in the narratives. This led to the possibility that, although by no ordinary means could the subjects have become aware of these details, they might somehow, under hypnosis, be picking the details up by clairvoyance from the checklist, or by telepathy from the mind of one of the investigators. Alternatively, they could have been exchanging telepathic information among one another: or, even more bizarrely, they could have been picking up the information from the mind of one of the 'real' abductees. But any of these explanations depends on a communication process whose existence remains unproven, and so must be highly conjectural in any case; moreover, there is the question of motivation. There is a well-known tendency among hypnotic subjects to try to 'please' the hypnotist by giving him the answers he hopes for, and so it is reasonable to conjecture that in this case the hypnotized subjects would seek any means they could of feeding the investigators with the sort of narrative they wanted. Consequently the possibility that all the subjects had resorted to ESP in order to obtain the appropriate detail from which they could fabricate a satisfying narrative is one which, however improbable, we cannot leave out of account.

Yet even this would not make it any the less significant from the point of view of our study. For the point clearly emerges, that these subjects, who had certainly *NOT* had an abduction experience (unless we are prepared to countenance the hypothesis that all those who offered themselves for the experiment were in fact people who, without knowing it, had had an abduction experience in real life!), were able to obtain, and structure into a reasonably coherent narrative, detail of a kind that makes their accounts virtually indistinguishable from those furnished by supposedly 'real' abductees.

From this one of the inevitable deductions is that if the volunteer subjects could 'get hold of' such information, the same was true of the 'real' ones. All could be employing the same process to obtain the information: all the accounts could be equally imaginary.

But this is only one of several possible processes which may have taken place. For one thing, it leaves unanswered the question, where did the *first* supposed abductee get his information from? In principle, there would have to have been a primary source to serve as inspiration for those that followed; this could have been imaginary or fictional, and so again various explanations are available.

The evidence that the information was obtained at least in part from the minds of the investigators was certainly persuasive. In one instance there was apparently a direct quotation from a book which one of the investigators had been reading the very day of the experiment. It seems unlikely, however, that *all* the information was obtained from the same source: if we hypothesize that it was obtained by ESP of some kind, then it would have to have been the result of 'scanning' all available sources of information in order to accumulate a suitable collection of details. No single source could contain such varied information as:

★ smoke trails observed behind the UFO when seen in the sky

★ changes in the shape and size of the UFO

★ levitation of the witness into the UFO

★ floating sensation when the witness was moving

★ calming effect exerted by the captors

★ breathing apparatus used by the aliens

★ probing of witness' mind by the abductors

★ samples of body fluid taken from the witness

★ examination of witness' body, especially sex organs

★ examination by means of beam of light

★ telepathic communication

★ absence of marks after examination

And yet these and many other such details occur spontaneously in the imaginary accounts given by people who supposedly had never read or heard of such encounters.

Thanks to the fresh insight provided by the Imaginary Contacts, DeHerrera was able to take a second look at the Brian Scott case. There now seemed grounds for discounting the validity of Brian's original testimony as the narrative of a real incident. Adding these findings to the discrepancies of the original account, together with Brian's overall unsatisfactoriness as a witness, there seemed a real likelihood that the entire experience may have been fabricated. By this is not, however, necessarily meant a *conscious* fabrication: rather, it is conceivable that Brian, for reasons which would require psychological examination, was in an altered state of consciousness, akin to that of the volunteer subjects of the experiment, and that he, like them, was mentally programmed to concoct such a scenario.

It emerged that, previous to his alleged encounter, Brian had been exposed to a certain amount of UFO material: he had read a number of UFO books, and had a tape-recording of the soundtrack from a film, which between them could well have contained the requisite raw material for his story. It even included such details as the Nazca 'spider' which he had tattooed on his arm, the Bolivian connection, and many of the actual terms, names and phrases, incorporated in the 'messages' he passed on from the supposed aliens. In other words, Brian's mind had had access to all he needed to fabricate his account, considerably more conveniently than had the imaginary abductees.

DeHerrera hypothesized that Brian may have had a genuine UFO experience, which nobody believed: that he thought its incredibility was due to its being located in a city area, so he relocated it in the desert, adding the necessary details to enhance its credibility. All this could have placed an emotional strain on him which could have led to the genuine paranormal occurrences, such as the balls of light which certainly seem to have been real enough for others to witness.

Among the follow-up incidents was an occasion when Brian, working in his garage at 3 a.m., became aware of a 'little man' who suddenly appeared, standing in the doorway of the garage.

'This man had normal features but was much shorter than I. He was old, I would imagine, though his face was not wrinkled. He was bald-headed on top but long white hair fell straight down from the centre of his head to his neck. He wore a one-piece metallic suit with a high collar in the back. He wore a loose belt with small patterns on it, and a large clear medallion around his neck. At first he didn't say anything. He just stood there. Then he opened his mouth and sounds seemed to be coming from everywhere. Then he said "From beyond all time, I am." Soon he stepped back a short distance. Then he compressed into a small orange ball of light and floated away through the air.'

The similarity to other types of entity sighting, such as the apparition to the widow on the mountain (page 78) is curious: it is noteworthy, too, that Brian himself stated that the figure was not completely solid — he was unquestionably an apparition of some kind. Was he simply a hallucination, or some kind of projection; and if the latter, did he originate in Brian's mind or from some external source?

John DeHerrera has also carried out a related piece of research which he describes as a 'hypnosis and dream study'[31] designed to test witness' recall and its reliability. Again, subjects were asked to volunteer for hypnosis: they were then asked to describe a scene from their distant past. He found that, though accurate in many respects, these accounts were far from being completely so. Events would be telescoped — for example, a car which figured in a particular incident was subsequently found not to have been acquired until years after the incident itself had occurred, or participants in a particular episode were shown not to have been acquainted with the subject at the time the episode was supposed to have occurred. On other occasions details were shown to be totally inaccurate. The implication is that, in its desire to present a nicely-rounded narrative, the subconscious mind does not restrict itself to the details of the remembered incident, but pads them out with bits and pieces from all over.

A different kind of finding from the same study relates to a specific characteristic of the UFO abduction accounts. DeHerrera found that witness descriptions of sensations during the course of the actual abduction — of floating, falling, etc. — are in fact commonly reported by hypnosis subjects and are by no means exclusive to this particular type of abduction scenario. He compared them to dream sensations of being unable to move, or of floating gently to the ground after jumping or falling from a high place. He concluded that these details, at least, could be attributed simply to the fact that the subject was in an altered state of consciousness, and did not specifically relate to the imaginary abduction scenario.

Similarly, he found that subjects reporting dreams, or under hypnosis, would frequently describe other figures as speaking without moving lips or mouth: so this too is a feature that is not unique to alien contact cases. When it occurs, it is generally assumed, both by the witness and by the investigators, that communication was taking place via telepathy: but this may simply be a rationalization of a seemingly incongruous observation.

The DeHerrera/Lawson/McCall findings are still the subject of much discussion: their importance is recognized, but the implications remain far from clear. To some, it seemed that they were sufficient to disprove the reality of *any* abduction accounts: if people *could* imagine stories such as these, it was argued then one must take the view that they *did*. But Lawson offered the objection that there is a quite different emotional feel to the allegedly 'real' cases than there is to the imaginary narratives, and he noted several aspects of the 'real' cases that distinguish them from the experimental ones:

★ the real cases occurred involuntarily

★ the witnesses were often frightened

★ a time lapse was reported

★ physical effects were noted in some cases

★ there were physiological effects on the witness

★ amnesia occurred

★ there were psychological after-effects such as nightmares

★ there were psychical manifestations and other emotional effects

That these differences are real is not in question: however, they can be accounted for otherwise. For instance, they can be attributed to the altered state of consciousness which was the equivalent — in the 'real' cases — of the hypnotic state of the volunteers. In other words, the differences noted by Lawson could relate not to the experience itself, but to the cause of that experience — which may have been, let us suppose, an emotional crisis of some kind. Since in the case of the volunteer subjects these circumstances did not exist, the effects were also absent. [88,89,90]

When Lawson presented the results of his experiment at the MUFON 1977 Symposium, he titled his presentation *'What can we learn from hypnosis of imaginary abductees?*[84] At the end of his presentation he had to confess that the answers were far from clear; what made them important was that — whatever the process involved — it requires a radical rethinking of our attitude towards abduction accounts. Hitherto, the tendency has been for a few to accept these accounts as genuine encounters, and for others to dismiss them as fantasies, however sincerely believed in by those who presented them. The consequence was that cultists welcomed the abductees while serious ufologists regarded them as irrelevant. DeHerrera and Lawson show that either extreme is invalid, and that every alleged account may contain material that merits study.

For the purpose of our present inquiry, *these experiments establish, beyond question, the ability of hypnotized subjects to replicate, not simply in broad outline but in intricate detail, scenarios to which they have not had access by any conventional means. In the state of hypnosis — and it is resonable to conjecture that other states may serve equally well — subjects appear able to obtain access to material by non-physical and non-sentiment means, and then to re-structure that material on a creative and selective basis, using it to construct a dramatic, circumstantial and persuasively coherent account.*

Such considerations help us to evaluate other curious cases, such as two reported during the very earliest years of the UFO era. In 1947 a British freelance psychical researcher, Harold Chibbett, had the idea of questioning, under hypnosis, a subject with whom he frequently worked, to see whether she could throw any light on the new phenomenon of 'flying saucers' that was currently making the headlines. To his astonishment and considerable alarm, Chibbett got from his subject a full-blown abduction scenario of a kind that was not to appear more generally for another two decades.

The subject described herself as being on an alien planet, whose occupants were distinctly hostile. She developed stigmata ('H6AQ'), which she alleged was a mark inscribed by one of the aliens with a kind of gun, and which would never leave her. It was her belief that she was on the planet Mars.

A few months later Chibbett carried out a further experiment with the same subject. This time she reported that she was actually on a 'flying saucer', located just outside the Earth's atmosphere. She had been sucked into the craft because there was radium in her body, and the material used in the construction of flying saucers, facillinite, is attracted to uranium. Significantly, this episode occurred while Chibbett was trying to mentally project her to a destination in California: the implication was that she had been, as it were, hijacked en route, sucked in by the metal fabric of the vehicle she encountered as her body — her astral body? — hurtled through the skies. The similarity to an out-of-the-body experience is strengthened by her report that the occupants of the spaceship were unaware of her presence on board their ship. She was able to describe them, however, and did so in terms which are classic for such encounters, though at that early date there was no way she could have come across any parallel in the literature for none had been published at this date and few if any such alleged incidents had ever occurred. She described one figure of more than human size, together with several small eggshaped bipeds with webbed feet: she was able to understand what they said to one another, though she could not account for this.

Chibbett, at a loss to make any rational comment on a case which still seems bizarre even after the wave of abduction cases which have been reported in the ensuing third of a century, refers to a case in Australia which also dates from the early days of the UFO era. In 1955 an Australian hypnotist, in Adelaide, had been treating a ten-year-old schoolgirl named Janet. In the course of one session, when he asked 'Where are you?', she answered 'In a flying saucer'. She then proceeded to give a detailed description of a trip, complete with landing, visiting a city and seeing people. Her description was very similar to that given by George Adamski, whose book [93] had been published not long before, and in response to suggestion she stated that her space beings were in contact with 'an important man' on Earth whom she later agreed to be Adamski. She supplied no new information, and there was nothing in her story that could not have been obtained from easily accessible sources; it would therefore seem as if in this case we have a very simple version of what later subjects were able to carry to much more elaborate lengths. [17]

To encounter alien beings in an abduction situation is a very extreme experience; we do not have to be psycho-analysts to recognize the kind of emotional significance it may have for the subject. What the DeHerrera/Lawson/McCall experiments show is that comparable processes seem able to occur in situations where emotional involvement is, if not non-existent, at any rate on a very low level. This has considerable implications for other phenomena in the field of our inquiry.

2.2   ENTITY FABRICATION IN HYPNOTIC AGE-REGRESSION

Age regression by means of hypnotic suggestion is both fashionable and controversial. On the one hand, bold claims are made for its value as therapy; on the other it is hailed as a unique source of information, the next best thing to travel in time. However, critics have pointed out that much of the material presented by the subject as recollected memories of the past is nothing of the sort.

Fortunately, for the purpose of our study we do not have to get involved in this controversy. There is, however, one secondary aspect of the age-regression process which is relevant. Because the act of submitting to that process is something the subject does voluntarily, I have included this phenomenon in Part Two rather than Part One, as a contrived rather than a spontaneous occurrence; however, the specific entity manifestation with which we are concerned is an involuntary affair, and this is at best a borderline case. Since, though, it is part of the thesis of this book that no sharp line can be drawn between spontaneous and experimental phenomena, this ambiguity is rather welcome than otherwise.

Entity fabrication occurs in the context of hypnotic regression in the way the subject accounts for the presence of the hypnotist, once he has been projected backwards in time. However acceptable his presence may be in the present day, it is a total anachronism once the subject has been taken back — whether truly or spuriously is not important — to a 'previous existence' as a Roman soldier or a medieval peasant. How does the subject cope with this potentially awkward situation?

Hilgard[70] found that subjects are remarkably well able to handle the situation, and show themselves quite capable of assimilating the alien presence of the hypnotist into their allegedly historical experience:

> Of those judged not to have displayed regression, practically all identified the hypnotist correctly. Among those who were judged to be regressed, just under a third correctly identified the hypnotist, whereas the others saw him as a stranger who had intruded into the regressed experience. One of my subjects, regressed to a birthday party in a park, saw me as the caretaker who was picking up paper on a pointed stick.

Analyzing his findings, Hilgard found that 71 per cent saw the hypnotist as someone other than himself: that is to say, they created a role for him to play which would be consistent with the rest of the allegedly remembered experience. Of these, 52 per cent saw him as 'a stranger', and 17 per cent as 'a friend of the family'.

Yet again, here we have evidence of the creative power of the subconscious mind. When the need arises, it is able to give face, clothing and even a complex identity to its fantasy creation — for, whether the supposedly recalled experience occurred or not, there can be no doubt that the hypnotist was not present on the original occasion now ostensibly being recalled. If, of course, the occasion was a real one, the question arises, was there a real caretaker present?; and if so, was the percipient simply 'borrowing' his identity for Hilgard to use?

The question of the state of mind of a hypnotized subject is still wide open to a variety of interpretations. There is on the whole a consensus that he is in an altered state of consciousness, characterized by an increasing suggestibility associated with a narrowing of his focus of attention onto the hypnotist. In the process he seems to become to a considerable extent oblivious of other external factors, though there is some evidence to suggest that this is only apparent, and that a hypnotized person is in fact taking in his full quota of sensory input, though not responding to it as he would in the waking state. (The same is true, of course, of people in the sleeping state, as is evidenced by responses which appear in disguise in dreams, where sensations of cold will result in appropriate dreams, and so forth.)

There is substantial evidence to suggest that in the hypnotic state our psi-abilities may be enhanced. The French investigators of hypnosis in the nineteenth century noted that their subjects often displayed clairvoyant abilities: in our own time it has been shown that hypnotized subjects enjoy enhanced powers of memory recall. Clearly, hypnosis acts as a kind of de-inhibitor, allowing access to material not normally available, whether it is memories that were never articulated by the conscious mind, or material that reaches the mind by alternative channels, overriding the limitations of the conscious mind.

However, these gains are not achieved without paying a price. Hypnotized subjects seem to be incapable of discriminating between the kinds of material they are handling. As John Vyvyan observed in his perceptive study, [163] 'a hypnotized person sometimes tells a clear and coherent story that is really composed of blended material from a variety of sources, and is usually unable to differentiate between them'.

There has been debate as to whether this represents a raising or a lowering of the mental/spiritual level: but the consequence for the investigator is clear enough — that we cannot believe all that a hypnotized subject tells us. However veridical it seems, an account given in the hypnotic state is a fabrication which *may* be made up only of true events, but may also contain a small or large admixture of fiction. And in the process of recruiting actors to play a part alongside the 'real' participants in the action, the creative subconscious performs an entity-creating operation which seems outwardly to be no different from other kinds of entity-seeing. And if the outward results are the same, so may be the internal processes by which those results were achieved.

2.3   HALLUCINATION UNDER CONTROL

The story of 'Ruth'[136] is the only case history known to me in which a subject, troubled by hallucinations, has learned to systematically subject those hallucinations to voluntary control. There may have been other cases that have not been reported; there may have been tentative attempts to do this that have not succeeded so well; there may have been cases in which this control was exerted without the percipient realizing what he was doing. Consequently, we cannot say to what extent the Ruth case is exceptional: but this much we can say, that if it happened once to one person, then the potentiality exists for it to happen again to other people. It could well be that a great many other people who experience hallucinations could be as successful as Ruth was in putting the creations of their unconscious minds under the control of their conscious minds.

Ruth, aged twenty-five and happily married with three small children, came to the psychiatrist Morton Schatzman for help with personality problems, which were accompanied by troublesome dreams and recurrent sightings of apparitions of her father who, inquiry revealed, had sexually assaulted her when she was ten.

The character of these apparitions was remarkable in several ways:

1   They were totally natural in appearance: the hallucination resembled the apparent in precise detail as to clothes, features, movements, expressions and so on. However, this related to the appearance of the apparent *as known to the percipient;* she had not seen her father for several years, and her apparition was like him as he had been, not as he was now (for he was still living, though in America while Ruth was in England). The percipient's unconscious was able to make some adjustments, such as letting the hair grow longer, but there was nothing that would imply clairvoyance, no external knowledge that was not available to her mind, nothing that she could not have found in her own knowledge and memories of him.

2   The apparition appeared to blend perfectly with the setting in which it was seen. It would utilize the furniture that was there, move round objects, seem to lift articles from tables, and so forth. There were one or two instances in which the hallucination did something that

would have been physically impossible for a real person, such as walking through a closed door, but these were very much the exception and one is tempted to say that they were lapses due to carelessness on the part of the 'producer'!

Generally, the hallucination would open a door, walk through it and close it behind itself; it would be reflected in mirrors and cast a shadow. If there were other persons in the room, it would interact with them, moving as though aware of their presence and location, seeming to listen and take note of what they said and did, and responding naturally to the percipient.

3   The apparition gave the appearance of speaking, and would seem to carry on sensible and logical conversations with the percipient in language that was characteristic of the apparent, using expressions, turns of phrase and mannerisms appropriate to the original. Sometimes it said things that were unfamiliar or unknown to the percipient, but though they might be unexpected, they were the sort of thing the apparent might have said. In other words, these communications, while demonstrating the seemingly autonomous character of the apparition, do not necessarily imply an external agent.

If there were other people in the room at the time of these conversations, they were not aware of them. But if the apparition was speaking at the same time as someone else, Ruth herself would find it hard to hear the other person.

4   Ruth had the impression of receiving other sensory impressions related to the apparition's actions. She could hear all the appropriate sounds — doors opening, cigarettes being lighted and so on. Even characteristic smells were evident — her father had a special smell which Ruth often sensed even when his apparition was not manifesting. If the apparition left the room via the door, Ruth might feel a draught while the door was 'open' (which of course it was not). If the apparition seemed to sit on her bed, she would feel it yield beneath its weight. But there were no permanent traces, as was established when she deliberately tested an apparition by asking it to put some toothpaste on her toothbrush. It obligingly did so: but later she found no sign of any paste.

5   As a consequence of Schatzman's very positive line of treatment, apparitions of other persons subsequently appeared. Sometimes they were deliberately evoked, sometimes they appeared spontaneously and to Ruth's conscious surprise. However, except for one enigmatic figure who was never identified, all the apparitions were of people known to her, and living.

6   Similarly under Schatzman's guidance, Ruth learned to exert control over the apparitions, to the extent that she could cause them to manifest and dismiss them at will. She could even cause them to manifest in the same room as their own originals, side by side! However, once brought into being, they acted autonomously, and Ruth had no

other control over them beyond the ability to make them go away.

7   Some of the hallucinations represented fears — particularly, of course, her father's: others were friendly, such as her best friend Becky's. Travelling away from her husband, she was able to summon up an apparition of him sufficiently convincing for her to enjoy a total illusion of sexual intercourse with him.

All the foregoing is consistent with the hypothesis that Ruth's hallucinations had no separate existence or external origin, but were projections emanating from her own mind. This is the generally accepted interpretation of such phenomena, and Ruth's experience gives us no grounds for hypothesizing any other process. But what her experience does do is substantially enlarge our knowledge of the process; it shows the creative capacity of the subconscious to be considerably greater than has ever been reported previously. This is demonstrated especially in the seemingly autonomous behaviour of the apparitions, who even when she had knowingly and deliberately called them into being, would then proceed to act without any apparent direction from her. She herself said of them, 'They've got personalities. I can't make them do what they don't want to do, or keep them from doing what they want to do. They're like real people.'

She was often taken by surprise by their appearance, or by what they said and did. Frequently this would result in argument and conflict of 'wills': thus one day she reported:

'I saw my father while I was in the bathtub. I was frightened, which he enjoyed. I felt terrible about his looking at me, but I didn't try to cover my body. Then I decided not to let him enjoy it. I asked him to get me a towel. I didn't really believe he would. "Hurry up, I'm cold," I said, as hateful as I could . . . I threw the soap at him. "You little bitch!" he said. He opened the door to walk out, and I felt a draft of air.'

An episode such as that, while not different in kind from many of the entity manifestations we have looked at in this study, has a unique vividness, intimacy, verisimilitude. Is Ruth's experience simply an enhanced form of the same kind of process involved in other entity-sightings, or is something else involved? Clearly, one of Schatzman's tasks was to eliminate any possibility that the apparitions might have physical existence, and a number of tests were carried out with this intention. For instance, an experiment was set up whereby Ruth would summon up an apparition and cause it to move in front of lights, either still or flashing. Under these conditions he was able to establish that her eyes were in fact seeing the light, even though Ruth herself insisted that the light was obscured when the passing apparition passed in front of it. Similarly, Ruth got an apparition to turn the light on and off, and described the room accordingly. But when, in a darkened room, she told Schatzman that the apparition had switched the light on, she was unable to read a book despite her insistence that the light really was on.

The act of calling up a hallucination seemed to make energy demands on the percipient; Ruth was frequently exhausted after a hallucinatory experience. This implies a physical process of some kind: if energy was used, it must have 'gone somewhere'. Another kind of confirmation that the ability had a physiological basis came from Schatzman's discovery that Ruth's ability was to some degree shared by other members of her family. Both her father and other relatives had had apparitional experiences; thus on one occasion she saw an apparition of her husband sitting in a chair, and asked her (real) father to see if someone was there. He looked and not only saw an apparition but recognized it as her husband. There are alternative explanations for this, of course, in terms of telepathy, but there was evidence that other apparitional experiences had occurred to him.

This prompted Schatzman to replicate Ruth's experience with other subjects, but his attempts were unsuccessful, whether with other members of the family or with people known to him who had reported spontaneous hallucinatory experiences. None of them was able to match Ruth's experience. However, on one occasion her husband saw an apparition of Ruth herself, looked at it and spoke to it for some time before realizing it was not her physical self. It is tempting to jump to the conclusion that this was a case of Ruth projecting an apparition of herself; but if so, this would represent a quite contrary procedure from her usual experiences, in which she and she alone was able to perceive the apparition. It does, however, raise the possibility that a subject possessing one kind of ability might also possess the other.

There are two further significant features of Ruth's experience which may help us to understand the process:

★ When recalling the sexual assault, Ruth went into a trance state to enhance her visualization, and in that state she was seemingly able to live the incident not only from her own viewpoint as a ten-year-old child, but also that of her father as he was at the time. She was able to identify strongly with his feelings and motivations and understand his urge. While this could be a projection from her own feelings and knowledge about the episode, there is also the possibility that she was able to 'enter into his mind' in some manner; if so, it seems likely that this is related to her astonishing ability to recreate his appearance and manner of behaviour with such verisimilitude.

★ Ruth was also able to call up an apparition of herself, with which she could hold conversations. It was not a mirror image, but a perfect facsimile. This 'Ruth 2' appeared to have knowledge of things which the everyday Ruth did not know, and also to know better than 'Ruth 1' what was best for her.

Although she did not reveal anything that the living Ruth could not have known, she did recall things that 'Ruth 1' had forgotten. In this sense she seems to have been a kind of superior version of Ruth, aloof from the day-to-day, not so closely involved in Ruth's troubles, endowed with greater perception and wisdom. The parallel with the super-being that manifests in so many cases of multiple personality, like those of Sally Beauchamp or Eve, is very marked, and it seems reasonable to assume that this is an alternative version of the same process, which we shall be studying in greater detail in a later chapter.

It also seems reasonable to suppose that, as appears to be the case in multiple personality cases, Ruth's hallucinations represented an attempt to externalize her problems — in the first negative phase, by giving substance to her fears or by creating a scapegoat for her sense of guilt: in the later and more positive phase, as the embodiment of her hopes of achieving balance and understanding. Was the entire affair created by her subconscious mind with a therapeutic purpose — was she, as it were, frightening herself into a state where she would have to take some critical action, such as going to consult the psychiatrist? Schatzman himself observes: 'Why she had chosen an apparition as a vehicle for manifesting her memories, wishes or insights, I did not know. Had I been less interested in her apparitional experiences, she might have paid less attention to them herself. By dramatizing her emotional problems in the person of the apparition, she had allowed me an opportune means for gaining access to them.'

It would seem, though, that Ruth was making use of the means available to her — exceptional, seemingly unique means, which happened to be very suitable for the purpose. Clearly she possessed this ability, and when the need arose she brought it into play. We would have to suppose that her clever subconscious mind was well aware of her latent ability, and took advantage of it with the same kind of opportunism as we have already suspected in other kinds of entity manifestation.

The alternative possibility exists, of course, that Ruth did not possess this ability, not even latently, until the crisis occurred in her emotional life, and that her subconscious mind then bestowed this gift on her. But if so, we have to ask why similar crises in other people's lives do not bring out similar talents? We also have to ask why Ruth continued to enjoy the ability to conjure up hallucinations even when her crisis had been satisfactorily resolved: even after the psychotherapy ended, she told Schatzman that when driving on her own she would sometimes 'put' an apparition in the seat beside her, just for company, or that when bored at a party she sometimes conversed silently with an apparition.

★   ★   ★

Clearly the case of Ruth is exceptional: her ability was rare if not unique, and perhaps not everyone, even if they possessed the resource, would be able to bring it under control. Had the case been known to the doctors who treated hallucination cases in the past, we should certainly have had some interesting comparative material: as it is, we shall have to wait to see if other cases arise in which, with the Schatzman findings as a precedent, similar faculties and processes can be diagnosed. But even though Ruth's is evidently an extreme case, this does not make it any the less relevant for our present study; for what it demonstrates in a high degree may be something that exists more generally in a lower degree in the psychological make-up of many, or even all, of us.

For example, a friend of mine told me how, having to make a long and demanding overnight motor journey across Spain in order to cross by ferry to Africa for an important appointment, he had become aware of someone sitting in the passenger seat beside him. He did not see this person visually nor did he receive any indication of its identity; but he was strongly aware of its presence and also of the reassurance the entity gave in the form of moral support to offset the rigour of the drive. Such an experience has clearly much in common with the imaginary companions and counsellors we looked at earlier; but equally are we not justified in hypothesizing that this experience was a slighter version of Ruth's; that my friend, like her, had subconsciously called into being an entity to help him cope with a situation, and that he too, with the kind of guidance that Ruth received from Schatzman, might have developed the ability to evoke his companion at will and create an entity which, while manifestly the creature of his own will, would have developed a capacity for independent action?

We shall see that there are other cases to which we can compare Ruth's experience, but hers differs from theirs in one very specific respect: that the original of her entity was a real, living person. In this respect it seems closer to apparitions of the living — but without implying the external agency that most such cases imply. Clearly, Ruth's case presents a unique combination of features, and as such is a unique case forming a unique category. For this reason we must be careful not to place too much dependence on it. But insofar as it demonstrates in a particularly graphic form a process that may be operating more unobtrusively in many of our entity-seeing categories, it is arguably the most illuminating and suggestive case we have studied so far.

2.4 EXPERIMENTS IN PROJECTION

Apparitions of the dead or dying are customarily accounted for by supposing that the apparent, at the moment of crisis or tormented by earthbound feelings, acts as his own agent and somehow projects an apparition of himself to where the percipient sees him. To what extent this is a conscious, deliberate and controlled process can only be conjectured. Apparitions of *living* persons, on the other hand, would seem to be wholly unconscious: the apparent has no notion that a visual facsimile of himself is at large. If he is his own agent, then it must be his subconscious self — whatever that may be — that is responsible.

Such conclusions have naturally prompted the question: can what happens unconsciously and spontaneously be made to happen consciously and deliberately? The surprising thing is that the attempt has been made comparatively rarely, for here is an experiment that requires no costly apparatus, nor any particular skill, and is therefore within the reach of any amateur; yet it is capable of being tested by comparing the accounts of the agent and the percipient with hardly less precision than, say, a ganzfeld experiment in extra-sensory communication.

Nevertheless, a sufficient number of cases have been reported to establish voluntary projection as a phenomenon deserving our attention. Few were carried out under quite the rigorous conditions we might have preferred; but at the same time there is no reason to suspect the protagonists of deception, and to suppose it in every case seems unreasonable. Because these reports tend to be widely scattered in the literature, and because each case presents features of its own which help to define the phenomenon, I have thought it best to quote a number of specific cases, all of which seem to me both genuine and interesting.

1808: Wesermann

One of the earliest experimenters in any field of the paranormal was an enterprising German named Wesermann, who on several occasions successfully 'appeared' to people in their dreams or conveyed images to them in a primitive form of the ESP experiments of our own day: four of his subjects reported dreams in which they saw more or less what he willed them to dream.

In a fifth instance, things went interestingly wrong. The intention had been that Lieutenant N. should see, in a dream, at 11 p.m., a lady who had been dead for five years, who would incite him to a good action. It was supposed that at that hour the officer would be in bed asleep and dreaming; but in fact that night he sat up with a friend, discussing the campaign against the French.

Suddenly the door of the room opened, and the lady, dressed in white, with a black kerchief and a bare head, walked in, saluted N.'s friend three times with her head in a friendly way, turned to N., nodded to him, and then returned through the door. The two men rushed out after her, and questioned the sentinel at the entrance, but he had seen nothing. [61]

The other officer told Wesermann that the door of the room always creaked when opened, but that it had not done so on this occasion; the suggestion being that the opening of the door was as imaginary as the apparition itself. The case is unique in that the agent did not project his own image, but that of someone else; and the fact that there were two percipients is also unusual. There are, of course, many questions we would want to clear up before accepting this case at its face value; but in most respects it is consistent with other cases we shall be considering.

1855: Moule

During the exciting early years of spiritualism, John Moule, who was a practitioner of what was then called mesmerism, reported the following experiment:

> I felt very anxious to try and affect the most sensitive of my mesmerism subjects away from my house, and unknown to them. I chose for this purpose a young lady, a Miss Drasey, and stated that some day I intended to visit her wherever she might be, although the place might be unknown to me; and told her, if anything particular should occur, to note the time . . . One day about two months later (I not having seen her in the interval) I was by myself in my factory at Mile End, all alone, and I determined to try the experiment, the lady being in Dalston, about three miles off. I stood up, raised my hands, and willed to act upon the lady. I soon felt that I had expended energy. I immediately sat down in a chair and went to sleep. I then saw, in a dream, my friend coming down the kitchen stairs, where I dreamt I was. She saw me, and suddenly exclaimed 'Oh, Mr Moule!' and fainted away. This I dreamt, and then awoke.
>
> I thought very little about it, supposing I had had an ordinary dream; but about three weeks after, she came to my house, and related to my wife the singular occurrence of her seeing me sitting in the kitchen, where she then was, and that she fainted away, and nearly dropped some dishes she had in her hands. All this I saw exactly in my dream, so that I described the kitchen furniture, and where I sat, as perfectly as if I had been there, though I had never been in the house.

The positive effort made by the agent is a feature that we shall see recurring, though we must not necessarily assume that such an expenditure of energy is really required. It may simply be that the agent has to put himself into a particular state of mind, and that the effort of will, such as Moule made, is one way of concentrating the mind and perhaps inducing the necessary state — in other words, it may be not so much the expenditure of force as the focussing of attention that is the important part. It is noteworthy that though Moule was standing when he made his effort, and sat down later to recover, it was as a seated figure that Miss Drasey saw him.

The case is particularly notable in that the incident was seen by the agent simultaneously. He did not know, of course, that his experiment had been successful until it was later confirmed; in other cases we shall see that agents, though they may have felt themselves to have been successful, had similarly to await confirmation from the percipient. Also, in this case, the question of expectation on the percipient's part can virtually be discounted, since several weeks had elapsed since the project had been mentioned.

1859: Russell

A Mrs J.M. Russell reported to the SPR:

> I was living in Scotland, my mother and sisters in Germany. On a sudden I made up my mind to go and see my family. They knew nothing of my intention; I had never gone in early spring before; and I had no time to let them know by letter that I was going to set off. I did not like to send a telegram, for fear of frightening my mother.
>
> The thought came to me to will with all my might to appear to one of my sisters, never mind which, in order to give them a warning of my coming. I only thought most intensely for a few minutes of them, wishing with all my might to be seen by one of them — half-present, in vision, at home. I did not take more than ten minutes, I think. I wished to appear at home about 6 o'clock that Saturday.
>
> I arrived at home about 6 o'clock on the Tuesday morning following. I entered the house without anyone seeing me. One of my sisters turned round when she heard the door opening, and on seeing me, stared at me, turning deadly pale and letting what she had in her hand fall. I spoke, 'Why do you look so frightened?' She answered, 'I thought I saw you again as Stinchen [another sister] saw you on Saturday.'
>
> She told me that on the Saturday evening, about 6 o'clock, my sister saw me quite clearly entering the room in which she was by one door, passing through it, opening the door of another room where my mother was, and shutting the door behind me . . . They looked everywhere for me, but of course did not find me. My mother was very miserable; she thought I might be dying. [61]

1881: Beard

A Mr S.H. Beard reported to the SPR:

> On a certain Sunday evening, having been reading of the great power which
> the human will is capable of exercising, I determined with the whole force
> of my being that I would be present in spirit in the front bedroom on the
> second floor of a house situated at 22 Hogarth Road, Kensington, in which
> room slept two ladies of my acquaintance, viz. Miss L. S. Verity, and Miss
> E. C. Verity, aged respectively twenty-five and eleven years. I was living
> at this time at 23 Kildare Gardens, a distance of about three miles from
> Hogarth Road, and I had not mentioned in any way my intention of trying
> this experiment to either of the above ladies, for the simple reason that it
> was only on retiring to rest upon this Sunday night that I made up my mind
> to do so. The time at which I determined I would be there was 1 o'clock in
> the morning, and I also had a strong intention of making my presence
> perceptible.
>
> On the following Thursday I went to see the ladies in question, and in
> the course of conversation (without any allusion to the subject on my part)
> the elder one told me that, on the previous Sunday night, she had been much
> terrified by perceiving me standing by her bedside, and that she screamed
> when the apparition advanced towards her, and awoke her little sister, who
> saw me also . . . Besides exercising my power of volition very strongly, I
> put forth an effort which I cannot find words to describe. I was conscious
> of a mysterious influence of some sort permeating in my body, and had a
> distinct impression that I was exercising some force with which I had been
> hitherto unacquainted. [61]

Beard noted that both the sisters saw the apparition dressed in the
same way, and standing on the same spot. The gas was burning very
low, yet the phantasmal figure was seen with far greater clarity than
a real person would have been under those conditions. As with both
our previous cases, Beard's apparition was able to adapt itself to the
environment in which it 'found itself': even if he had been familiar with
the girls' bedroom, he would still have to position himself with regard
to the furniture.

A year or so later he tried again, and was seen by his percipients on
two separate occasions. But there were some differences. On the first
of these occasions he was in what he himself describes as 'a mesmeric
sleep' — a self-induced hypnotic state, in which he was conscious, but
incapable of movement. On the second occasion he was asleep. There
does not seem to have been any intention on his part of carrying out
the actions which were then reported by the percipient, who in this case
was not at all well acquainted with Beard; she had met him only once,
and had forgotten what he looked like, though on seeing him so distinctly
on this occasion she recognized him at once:

> About 12 o'clock I was still awake, and the door opened and Mr Beard came
> into the room and walked round to the bedside, and there stood with one
> foot on the ground and the other knee resting on a chair. He then took my

hair into his hand, after which he took my hand in his, and looked very intently into the palm. Ah, I said, speaking to him, you need not look at the lines, for I never had any trouble. I then awoke my sister.

But in this case a complicating factor intrudes: the sister did not see the apparition! How can we account for this? It is probably significant that in the earlier case Beard knew that there were two sisters living in the house, and may even have known that they slept in the same room; in this later case, however, the percipient's sister was just visiting, and Beard may not even have known of her existence, let alone that she was sleeping in the house.

In this case, there is almost no doubt that Beard did *not* know the percipient's bedroom — indeed, it seems unlikely that he even knew where the house was. Such cases make it clear that if the apparent is also the agent, it is not acting by remote control, as it were, from the mind of the agent sitting or lying at home: it possesses sufficient autonomy to regulate its movements in accord with the actual topography it encounters in the course of the visit. This does not necessarily mean that the projected entity has been granted autonomous powers, given a free hand by the agent's controlling mind: it could be that the projected entity is able to send back information to the mind, which can then control its movements. We have no reason to suppose that such a process is affected by time and place, and we have no right to make any such assumptions at this stage.

Nevertheless, the fact that the apparition was seen only by the person Beard was thinking of, and not by someone else who was certainly in a position to see it, is a formidable complication. It implies either that the percipient had a special gift her sister lacked — which is an assumption for which we have no grounds whatever — or that Beard's apparition was itself able to control who saw it and who did not. By far the most logical deduction, in short, is that the intention of making oneself seen by a particular person — whether that intention is conscious or unconscious — is a fundamental part of the process. But not all the decisions can be made in advance; the business of the hand-reading, for example, seems to have been a spontaneous action on the apparition's part. It would be interesting to know whether Beard, in the course of his daily life, made a practice of looking at people's hands, and whether he had at any time contemplated reading that of his percipient on this occasion.

1888: Backman

A Swedish researcher, Dr Alfred Backman, carried out a series of experiments in which he got hypnotic subjects to pay 'mental visits' to places and describe what they saw. He obtained some very interesting results from these early 'remote viewing' experiments, and on one occasion a remarkable incident occurred. He asked his subject, Alma, in the hypnotic state, to visit a place she had never been to, the home of the Director-General of Pilotage in Stockholm, and see if he was at home. She reported that he was sitting writing at a table in his study. Amongst the details she mentioned was a bunch of keys beside him on the table. Backman told her to shake the keys, and attract the man's attention by putting her hand on his shoulder. This was repeated two or three times, and Alma was sure the man was aware of her presence.

The officer, who had no idea that the experiment was being attempted, confirmed that he had been sitting, fully occupied with his work, when, for no reason, his eyes fell on the bunch of keys on his table. He then started wondering why he had put the keys there, since it was not his habit to do so. While occupied with these thoughts, he caught a glimpse of a woman. Thinking it was a maidservant, he took no notice; but when he became aware of her a second time, he called her and got up to see what was the matter. But he found nobody, and was told that neither the servant nor anyone else had been in his room.

It is noteworthy that, though his attention was certainly directed to the keys, which would seem to be in accordance with the agent's wishes, he did not see them move or hear them rattle. [148d]

1890: Moses

In this case, the identity of the agent is unknown; the well-known spiritualist Stainton Moses was the percipient. The anonymous agent reported an experiment carried out about 1890:

> One evening early last year, I resolved to try to appear to [Mr Moses] at some miles distance. I did not inform him beforehand of the intended experiment, but retired to rest shortly before midnight, with thoughts intently fixed on him, with whose room and surroundings, however, I was quite unacquainted. I soon fell asleep, and awoke next morning quite unconscious of anything having taken place. On seeing him a few days afterwards, I inquired, Did anything happen at your rooms on Saturday night? Yes, replied he, a great deal happened. I had been sitting over the fire with M., smoking and chatting. About 12.30 he rose to leave, and I let him out myself. I returned to the fire to finish my pipe, when I saw you sitting in the chair just vacated by him. I looked intently at you, and then took up a newspaper to assure myself I was not dreaming, but on laying it down I saw you still there. While I gazed without speaking, you faded away. Though I imagined you must be fast asleep in bed at that hour, yet you appeared dressed in your ordinary garments.
>
> A few weeks later the experiment was repeated with equal success, I, as before, not informing him when it was made. On this occasion he not only questioned me on the subject which was at that time under very warm discussion between us, but detained me by the exercise of his will some time after I had intimated a desire to leave. This fact, when it came to be communicated to me, seemed to account for the violent and somewhat peculiar headache which marked the morning following the experiment. [61]

These experiments, if we can believe the account, are remarkable to the point of uniqueness. A certain doubt creeps in when we recall that Moses was not only a known psychic but also a determined spiritualist, which would make it possible for us to devise some kind of explanation related to his psychic faculties. This would affect the process without diminishing the fact: however, it would also place the responsibility for the incident at least as much on the percipient as on the agent, and this does not seem consistent with our other cases. What is more acceptable is the hypothesis that Moses, on experiencing the sighting, was better equipped to handle it than most percipients, and was able to make more of it, even to the extent of exercising some control over what happened by directing the course of the conversation and retaining the 'visitor' even when it was anxious to leave.

Unfortunately, only Moses seems to have had any recollection of the conversation. If it really occurred, of course, it adds greatly to the interest of the case, for it implies that the projection was not simply of the agent's physical appearance, but also of his intellectual self — the 'real' him, we would be entitled to say. I am inclined to think, however, that the projection may have consisted only of the agent's physical form; and that it was Moses who created the conversation, dramatizing the visit and putting words into his friend's mouth, just as Ruth, in an earlier chapter, put appropriate remarks into her father's apparition's mouth.

The overriding significance of such a case is, of course, that two people are involved, and both actively. It has many of the characteristics of a classic apparition case — the appearance in a strange room, the suiting of the manifestation to the exact lay-out of the room, the appropriate behaviour towards the percipient — that we found it so hard to account for when considering apparitions: but now we have the extra dimension, that whereas in the case of the apparition we could only conjecture as to whether the apparent was also the agent, in this case there is not the slightest doubt: the incident occurred only because the apparent/agent caused it to occur.

1955: Von Szalay-Bayless

Not only most, but all the *best* cases of voluntary entity-projection seem to have occurred a hundred years or so ago; however the following case, though not particularly dramatic, shows that such incidents continue to occur.

The American investigator Raymond Bayless reported that, on 5 February 1955, he was sitting in his bedroom, performing the mundane action of tying his shoelaces, when 'I saw a flickering motion before me, and I looked away thinking it was my cat, looked again, and saw a curious shadow in front of me which was trapezoidal in shape. Roughly the height of a man, it leaned from the perpendicular at an angle of about 25 degrees to my right, and "floated" in the air. It had a space of about one foot between it and the floor.'

Bayless watched it in astonishment: after a moment, however, the shadow rushed out of the room, through an open doorway, and disappeared. Later that morning he called on a friend, Mr von Szalay, and had said no more than 'Guess what happened to me . . .' before his friend interrupted him to reply 'You saw me!', and described how he had deliberately projected himself to Bayless' house. [6]

This by no means exhausts the list of such cases, most of which were on similar lines without any special features. Some contain suggestive details, nevertheless. For instance, in 1890 a Mr Kirk, of Plumstead, tried on several occasions to project himself to the home of Miss G., always doing so late at night. On one occasion he succeeded sufficiently well to disturb his friend, who felt so uneasy that she got up and did some needlework though it was 2 o'clock in the morning. When he eventually did make a successful projection, however, it was in the middle of the afternoon when, exhausted by some arduous office work, he was resting in his office and decided to make an experiment on the spur of the moment. As it happened, Miss G. was herself resting at that unusual time; she awoke and saw Kirk clearly, dressed as he then was. Noteworthy, too, was the way the apparition adjusted itself to the room his friend was in; she reported that 'he passed across the room towards the door, which is opposite the window, the space between being 15 feet, the furniture so arranged as to leave just that centre clear; but when he got about 4 feet from the door, he disappeared'. [148f]

Clearly related *somehow* to these phenomena, though it is far from easy to say how, are those cases known as 'bi-location' — where someone appears to be in two places at the same time. This is one of the classic phenomena associated with religious mystics: until the scientific investigation of out-of-the-body phenomena, it was regarded as a miracle, a sign of grace. Today we would more likely be apt to classify it as an unusual psychic ability, which does not make it any the less remarkable, but does enable us to study it along with other related phenomena.

The cases of bi-location told of the saints and others are not, of course, experimental like the cases we have been considering in this chapter. Strictly speaking, they are spontaneous incidents and should have been discussed in Part One of our study. However, as we have already seen, there are many instances in which the borderline between voluntary and involuntary is not hard-and-fast, but a matter of interpretation; and to distinguish between the conscious will of an experimental projector, and the subconscious will of a person who needs to be in two places at the same time, is begging a question that we shall have to confront sooner or later.

I do not propose to cite examples from the classic literature of religious mysticism, though it is here that the majority of such accounts are to be found. Though, as Thurston[159] acknowledges, the evidence for them is often very strong, it is also remote in time, and understandably we are apt to be suspicious of testimony that derives only from those who have an interest in establishing the veridicality of the stories. Fortunately, there are also some more recent instances, of which one of the most interesting is that of Natuzza Evolo, a housewife of Calabria, Italy, who was born in 1924.

Throughout her life Natuzza has displayed a wide range of para-normal phenomena, including one which is to the best of my knowledge unique, haemographs, in which the blood she exudes from stigmatic wounds is transferred onto a handkerchief or some other article, forming pictures or words; this has been observed by witnesses who certainly seem trustworthy.

One night in 1955 a Sr. Naquaniti, an acquaintance of Natuzza's, was wakened from sleep by loud bangings on the balcony. Suspecting a burglar, he went for his gun, while his wife, frightened, fled towards the lower part of the house, only to be confronted in the doorway by the apparition of Natuzza, accompanied by the figures of her dead father and uncle. The apparition smiled at her, then all three vanished. Though they did not mention it to Natuzza, she herself mentioned it next time they met, and joked with Sr. Naquaniti about going for her with a gun! All the time, of course, she was safely at home.

She herself says that these bi-locations do not occur as a result of her spontaneous will: instead, angels or spirits of the dead present themselves to her and take her to the place where her presence is required: 'Here I can see the place where I am so that I can describe it, and I can speak and be heard by the people present. I can open and close doors, move objects and produce actions. My vision is not a distant one, like watching a film or television, but I am plunged into the surroundings . . .'

At the same time, she is fully aware that this is not her physical body, which she knows she has left at home. The excursions may occur when she is asleep or awake: it would be interesting to know what state she is in when bi-locating during the waking state, but this does not seem to have been noted.

Though Natuzza's experiences have much in common with classic out-of-the-body events, there are also some differences. She has no sense of travelling — the translocation is instantaneous. The experience does not make her feel ill or tired. And of course the fact that her presence is sensed, and occasionally seen, is unusual and significant.[421]

Another extraordinary case was reported to the SPR in 1963 by a Mr

Lucian Landau, whose wife Eileen had frequent out-of-the-body experiences. The particular case that concerns our inquiry dates from 1955, at a time when the two were not married, but when Eileen was visiting Lucian Landau and sleeping in the spare bedroom.

Mr Landau was not well at the time, and one morning Eileen told him that during the previous night she had visited him — in her astral, not her physical body — to check his pulse and breathing. He asked her to do it again the following night, and if possible to bring some object with her: he suggested his diary, which she put on a desk in her room.

The next morning he awoke: it was about dawn, and enough light was coming through the partly-drawn curtains to read by. He saw the figure of Eileen, wearing a night dress and looking unusually pale, facing the window of his room, and moving slowly backwards though without actually walking; he calculated her rate of progress to be about one metre in fifteen seconds. He got up, and followed the figure through the doorway — which had been purposely left open — onto the landing. From this position he could see not only the figure but also the real Eileen, asleep in her bed, the bedclothes rising and falling as she breathed.

The figure continued to retreat, finally vanishing when it was more than halfway across Eileen's bedroom. Returning to his own room, Lucian found a rubber toy dog, which belonged to Eileen, lying on the floor near his bed; next morning Eileen said that she had not been able to pick up the diary, perhaps because it was not hers, and so had brought her toy dog instead. (Eileen suggests that her inability to pick up the diary may have been because she was brought up not to handle other people's letters or diaries.) She also said that, during her experience, 'my consciousness appeared to be normal, and so did my ability to see my surroundings. I do not remember anything about going backwards to my room, or entering my bed', but she admitted 'I felt very tired and wanted to go back to bed'.[147e]

This story is one of the most remarkable cases in the literature; it is unfortunate that it depends entirely on personal testimony, without confirmation, and without any similar case having been reported elsewhere. It is, however, consistent with our other cases, and we have no reason to doubt the two witnesses' word.

Analagous to these experimental cases are those in which two people make a compact that, after death, the dead one will attempt to visit the survivor — this being regarded as the surest means of demonstrating survival after death. There are, of course, objections that make it something less than watertight proof; fortunately, it is not with this aspect of such cases that we are concerned here.

What distinguishes compact cases from those we have been looking at earlier in this chapter is, of course, the fact that the experiment — assuming that what occurs is an experiment — is carried out by a dead person: consequently we cannot be sure what is happening, or what is intended, and we are free to suppose that there may be some alternative explanation. Indeed, from one point of view it seems more logical to class it along with the apparition cases we considered in Part One, where there is good reason to suppose that some kind of external agent is at least partly responsible for the manifestation. However, we cannot simply class compact cases as spontaneous apparitions either, for the matter of the compact is after all the crucial factor. Whatever the mechanics of the thing, we must assume that it would not have occurred at all unless the compact had been made.

There are quite a number of such cases in the literature; here are two, of which the first is a very old case which shows the phenomenon at its simplest and most archetypal.

Mr Smellie, author of *The Philosophy of Natural History*, made a solemn compact with the Revd William Greenlaw that the first to die should return and visit the other; but if the deceased did not appear within a year, it was to be concluded that he could not return. Greenlaw died on 26 June 1774. As the anniversary approached without a sign from his friend, Smellie grew extremely anxious: but eventually, just before time was up, Greenlaw appeared to him while he slept in an armchair, and told him that he had had great difficulty in getting through.[115]

The most celebrated of such cases is that narrated by the eminent statesman Lord Brougham who in 1799, at the age of twenty-one, was journeying with friends in Sweden:

'Arriving at a decent inn, we decided to stop for the night. I was glad to take advantage of a hot bath before I turned in, and here a most remarkable thing happened to me — so remarkable that I must tell the story from the beginning.

After I left the High School, I went with G., my most intimate friend, to attend the classes in the University . . . We frequently discussed and speculated upon many grave subjects — among others, on the immortality of the soul, and on a future state. This question, and the possibility, I will not say of ghosts walking, but of the dead appearing to the living, were subjects of much speculation: and we actually committed the folly of drawing up an agreement, written in our blood, to the effect that whichever of us died the first should appear to the other, and thus solve any doubts we had entertained of the 'life after death'. After we had finished our classes at the college, G. went to India, having got an appointment there in the Civil Service. He seldom wrote to me, and after a lapse of a few years I had almost forgotten him; moreover, his family having little connection with Edinburgh, I seldom saw or heard anything, and I had nearly forgotten his existence.

I had taken a warm bath, and while lying in it and enjoying the comfort of the heat, after the late freezing I had undergone, I turned my head round, looking toward the chair on which I had deposited my clothes, as I was about to get out of the bath. On the chair sat G., looking calmly at me.

How I got out of the bath I know not, but on recovering my senses I found myself sprawling on the floor. The apparition, or whatever it was, that had taken the likeness of G., had disappeared . . .'

Subsequently, Brougham, on his return to Edinburgh, learnt from a letter that G. had died on 19 December, the date of the apparition.[61]

While cases such as the foregoing ostensibly indicate that an effort has been successfully made by the dead person to manifest to the living, there remains the alternative possibility, that the percipient has received extra-sensory information as to the other person's death, and has projected the apparition from his own memory, building it into the dramatic form of a fulfilment of a compact. However, this explanation, which is in any case far from plausible, could only with the greatest ingenuity be made to account for a case such as the following.

The Revd Arthur Bellamy's wife had at school made an agreement with a fellow pupil, Miss W., that whoever died first would revisit the other. News of the former schoolgirl's death came casually in 1878, though many years had elapsed without contact; the compact was recalled.

Mr Bellamy stated:

A night or two afterwards, as I was sleeping with my wife, a fire brightly burning in the room and a candle alight, I suddenly awoke and saw a lady sitting by the side of the bed where my wife was sleeping soundly. At once I sat up in the bed, and gazed so intently that even now I can recall her form and features. Had I the pencil and the brush of a Millais, I could transfer to canvas an exact likeness of the ghostly visitant. I remember that I was much struck, as I looked intently at her, with the careful arrangement of her coiffure, every single hair being most carefully brushed down.

How long I sat and gazed I cannot say, but directly the apparition ceased to be, I got out of bed to see if any of my wife's garments had by any means optically deluded me. I found nothing in the line of vision but a bare wall. Hallucination on my part I rejected as out of the question, and I doubted not but that I had really seen an apparition.

Returning to bed, I lay till my wife some time after awoke, and then I gave her an account of her friend's appearance, described her colour, form, etc., all of which exactly tallied with my wife's recollection of Miss W. Finally I asked, 'But was there any special point to strike one in her appearance?' 'Yes,' my wife promptly replied, 'we girls used to tease her at school for devoting so much time to the arrangement of her hair.' It was the very thing which so much struck me.[108]

Although the majority of the cases I have quoted in this chapter occurred a hundred or more years ago, they seem in every case to have been conducted and observed by responsible people, well aware of what they were doing and of the implications of what they achieved. It is surprising that there has not been further experimentation along these lines: nonetheless, there are a sufficient number of successful cases to give us good grounds for supposing that this is a real capability, for which the mechanism must somehow exist within the make-up of quite a number of people — for it is notable that, for the most part, the agents in these cases were not particularly psychic or in any other way remarkable, but seem to have been more or less normal people motivated by simple curiosity. This being so, it is legitimate to wonder whether the same ability that we see here operating in experimental cases, may also be operating, at a subconscious level, in at least some categories of spontaneous case.

We shall be drawing heavily on these experimental cases when we come to make our final analysis, at which point we shall be concerned with what they have in common with spontaneous sightings. For the moment, let us simply note what seem to be the necessary conditions and characteristics of the experimental projection:

★ The agent, though not necessarily in an altered state of consciousness, is usually so: often he is actually asleep. It could well be that something of the sort is true also of those cases where we do not know anything about the state of mind of the agent, for there are several instances in which he speaks of a very determined effort of will, and this of itself may be sufficient of an altered state to provide the necessary 'climate'.

★ There are some cases which suggest that the entity is seen only by those whom the agent has specifically in mind: if someone else is present, without the agent's knowledge, that other person may not see the entity. We should note, though, that this is not true of the compact cases — to judge, at any rate, by the Bellamy case — nor of the Wesermann case with which this chapter opened.

★ In every case without exception, however, the entity acts in a way that implies awareness of the surroundings in which it finds itself: it makes use of doors, it skirts furniture, it knows where the percipient is and behaves as though aware of his location and often of his presence.

★ Expectation on the part of the percipient seems to be in no way necessary. In some of our cases, the percipient knew in a general way that some such experiment was being contemplated, but in none of them was any prior arrangement made to suggest that the effort was a co-operative one. Nor was the percipient in any particular state of mind: though some were admittedly passive, frequently waking from sleep, there were others who were normally occupied, though generally in restful ways such as writing.

★ In virtually every case, the agent was not aware of the experience while it was occurring. There are two exceptions: the first took the form of a dream, which the agent did not at the time think to be in any way veridical; the second was a special case for the agent was being hypnotized by someone else to whom she was obliged to report back while carrying out her 'visit'.

Other features and implications of these experimental cases will emerge when we compare them with spontaneous cases. But even at this stage we are justified in saying that, unless we can come up with a better explanation, these experimental projections support the hypothesis that we can detach a 'portion' of ourselves and despatch it to a distant and unfamiliar destination, where it manifests with sufficient substance to be seen by others, and with sufficient awareness of its surroundings to move naturally in the physical environment and behave appropriately in respect of the percipient.

2.5 SUMMONING ENTITIES BY MAGIC

Today it is customary for apologists for magic to disclaim material aims; instead, they claim that the ultimate purpose of their practices is union with the infinite, or transcendence, or some such abstraction. Throughout history, however, the declared purpose of the magician has been nothing so ineffable. It has been the achievement of power; and one of the chief ways in which power was to be achieved was by securing the services of agents who, possessing superhuman power, could be constrained to use it on the magician's behalf. Such agents, so powerful in many ways, were nonetheless inherently inferior to mankind, or at least to those individuals who had acquired the necessary knowledge and possessed the necessary will.

Consequently the conjuring up of demonic forces, in some incarnate materialization, has always been one of the highest feats to which the magician aspired, and has figured prominently in the history of magical practice. As David Conway, one of the few coherent latter-day writers on the subject, states:

> To the successful occultist belongs the deep satisfaction derived from identifying himself with the macrocosm of which he is a tiny part. By attuning his mind to the universal forces around him, he is able to expand his consciousness until everything is contained within himself. This he does by inducing supernormal states of consciousness which enable him to contact forms of existence that are not usually part of our human experience. [20]

Even Conway uses imprecise terms, but in practice those 'forms of existence' are usually known as 'demons', and 'contact' signifies not so much the friendly confrontation the word usually implies — a meeting of equals — but something more like ringing the bell for a servant.

What are these demons? Theories vary widely. Within the Christian belief-system they tend to be seen as fallen angels, condemned for having abandoned the service of God and now occupying Hell. The Churches naturally disapprove of any human trafficking with these beings, and consequently condemn all magical practices.

This view of demons has coloured most thinking on the subject; they are almost universally seen as evil. The main task of the magician is to assert his will over that of the evil being, but this is made a little easier because most of the purposes for which he needs him are also evil, or considered to be so by society as a whole — obtaining wealth and earthly power by unscrupulous methods, making innocent girls do what he wants them to do, and so on. So, as with trainers of dogs and seals, the magician is exploiting pre-existing proclivities.

Nevertheless he does so only at a price. To begin with, he has to invest years of study and inner development in acquiring the necessary skills and will: and if he is unable to secure the demon's services without signing a contract with his heart's blood, he will find eventually that he pays at an exorbitant rate for his earthly pleasures. The Church holds that he will pay that price anyway, contract or no contract, for the mere practice of magic is a sin which he will have to expiate.

However, it is not necessary to accept this biased view; we do not have to think of demons as inherently evil. An alternative view of them is that they are impersonal forces, a sort of elemental being, neither good nor bad, waiting like servants at a hiring fair for someone to summon them up. It is the magician's evil inclinations — according to this interpretation — that makes it seem as though the spirits are evil; like Nazi concentration camp officials, they can claim that they are only carrying out orders.

Whatever their nature, if they exist, and if humans are able to summon them up, this is something of relevance for our study. There are theorists who believe that a great many of our categories of entities are demons: one school of thought has held that even the ghost is not a projection of the apparent, but a simulation created by demonic powers. This was the official Protestant view at the time of the Reformation, as expressed by Lavater[87] and others who were unable to accept the Catholic doctrine whereby spirits of the dead could return to Earth. Today there are theorists who believe that the UFOs are piloted by demonic beings[22, 169] and that their purpose is to work against the true faith in Jesus Christ. Indeed, the demons are used universally as a scapegoat, a convenient expedient for anyone who cannot believe that the inability of human beings to run their world properly is their own fault, but who insist that outside influences must be responsible.

It is not part of this study to comment on these ideas: we can only judge them by the results. The true nature of these demonic entities, which have as many shapes and forms as there are theories as to their nature, is unknown. There is, of course, a traditional appearance attributed to devils, including a number of physical signs whereby they can be recognized in cases of doubt; but we may reasonably doubt that this is their *true* appearance, even if they should turn out to be real creatures. It seems more likely that they take on a particular appearance as it suits them, or that we humans somehow invest them with what we expect to see. This is a profound question to which we shall return at a later stage: for the moment, let us occupy ourselves with the more obvious question of whether there is any evidence for their existence in the first place. Which, in the context of magic, is equivalent to asking,

What evidence is there for the literal conjuring up of demons by magicians?

Rather than get involved with the magicians of past ages, accounts of whose practices are so obscure as to be worthless, let us look at the work of a relatively modern magician, Aleister Crowley. In his introduction to the *Book of the Goetia*, he has this to say:

> I am not concerned to deny the objective reality of all 'magical' phenomena; if they are illusions, they are at least as real as many unquestioned facts of daily life; and, if we follow Herbert Spencer, they are at least evidence of some cause. Now, this fact is our base. What is the cause of my illusion of seeing a spirit in the triangle of Art? Every smatterer, every expert in psychology, will answer: 'That cause lies in your brain'.
> Magical phenomena, however, come under a special sub-class, since they are willed, and their cause is the series of 'real' phenomena called the operations of ceremonial Magic. These . . . produce unusual brain-changes: herein consists the reality of the operations and effects of ceremonial magic, those phenomena which appear to the magician himself, the appearance of the spirit, his conversation, possible shocks from imprudence, and so on, even to ecstasy on the one hand, and death or madness on the other.
> But can any of these effects . . . be obtained?[27]

I *think* Crowley's answer is 'Yes': probably the most literate writer in English on the subject, he could be admirably lucid when he wished — and infuriatingly evasive when he chose.

What follows, in that particular book, is a whole mess of scripts which tend to start 'I do invoke, conjure, and command thee, O thou spirit, to appear and to show thyself visibly unto me before this Circle in fair and comely shape, without any deformity or tortuosity, by this name of . . .', at which point the magician trots out a list of referees whose authority will add force to his commands.

Crowley was certainly one of those for whom the spirits are not inherently evil. He knew that the designs of the magician *could* often be evil, and that the spirits would have no choice but to serve those purposes; but it seems clear that he himself, for all his posturing, did not really think of himself as evil, nor the spirits he invoked. Those whom he lists in his instruction manual are all identified. The identification is of course handed down 'from antiquity', ancient catalogues which themselves derive from sources now lost. Crowley's 72 'mighty kings and princes' are attributed to the classification system of King Solomon. They are a mixed bunch, but one of the striking things about them is that most of them are not, as popular myth supposes, agents of destruction, but of *instruction*. For example Andrealphus, number 65 in the catalogue, can teach geometry perfectly; Cimejes, number 66, teaches grammar, while Amdusias, number 67, causes all manner of musical instruments to be heard, combining this with the capacity to make trees bend. Zagan, number 61, is one of several who make men witty — would that he were employed more often! — while Sitri, number 12, will, more mischievously, get women to show themselves naked if that is what the magician has in mind. There are of course a great many spirits who will undertake to cause women to love men, this being what a high proportion of magicians seem to have in mind; but in general Crowley's catalogue sounds more than anything else like the prospectus of an institute for self-development with the help of qualified instructors.

And the question still remains, was the summoning up of these beings ever actually *done?* Crowley himself not only hedges, but mocks his reader: he claimed to have sacrificed some 2000 male children of perfect innocence and high intelligence, a claim taken seriously by that most endearingly credulous of occultists, Montague Summers, who does not seem to have questioned Crowley's ability to lay his hands on so many exceptional children and abstract them without their parents noticing or objecting.

But there is better testimony than that. In 1909, along with one of his disciples, Victor Neuberg, who seems to have been as naive as he was gifted, Crowley went into the desert near Algiers and carried out a series of ritual performances. One of these was aimed at invoking Choronzon, who was for some reason omitted by King Solomon from his catalogue, but is well known, to the kind of people who know these things, as *The Dweller in the Abyss,* one of those splendidly mystical-sounding titles beloved of copywriters for the occult.

The invocation began with the sacrifice of three pigeons; then Neuberg placed himself inside a circle and Crowley inside a triangle marked on the hot sand: the idea was that Crowley should let himself be possessed by the demon. Crowley sat in his triangle dressed in a black robe, in a yoga position, and angels were invoked as guardians. The ceremonial invocation was then performed, and Choronzon arrived — not visibly at first, and using Crowley's voice, but nevertheless announcing his own arrival.

What Neuberg saw at this stage was not Crowley, but a beautiful courtesan he had loved in Paris. This he took to be a manifestation of, or caused by, the demon, so when the lady sought to lure him out of the protection of his circle, he prudently refrained. Then Choronzon himself took the girl's place, proving that Neuberg had been right to think she had only been a delusion; the spirit made a splendid speech, in which he offered to come into the circle and serve him. Again, Neuberg saw through the not very subtle subterfuge. Choronzon then proceeded to take on a variety of shapes, not willing to recognize that he was not likely to hoodwink the suspicious Neuberg with any of them; his most ingenious trick was to appear in the likeness of Crowley himself, without his robe, in which form he left his triangle and crawled across the sand towards Neuberg's circle, begging for a drink of water. Neuberg had the heart to refuse him, whereupon the demon reverted to his own likeness and tried a little braggadocio:

> I feed upon the names of the Most High. I churn them in my jaws, and I
> void them from my fundament. I fear not the power of the Pentagram, for
> I am the Master of the Triangle. My name is three hundred and thirty and
> three, and that is thrice one.

All of which was diligently taken down by Neuberg in shorthand.
Chastened, it seems, by Neuberg's adamant attitude, Choronzon
admitted his trickery:

> When I made myself like unto a beautiful woman, if thou hadst come to
> me, I would have rotted thy body with the pox and thy liver with the cancer.
> And if I had seduced thy pride, and thou hadst bidden me to come into the
> circle, I would have trampled thee under foot, and for a thousand years
> shouldst thou have been but one of the tape-worms that is in me. And if
> I had seduced thy pity, and thou hadst poured one drop of water without
> the circle, then would I have blasted thee with flame. But I was not able to
> prevail against thee.

But all the time he was chatting, and Neuberg was diligently scribbling
it down for the record, the scheming spirit was throwing sand onto the
protective circle, covering up the marks and thereby creating a chink
in the defences. Suddenly he leapt onto Neuberg, throwing him to the
ground. Fortunately, Neuberg had a magic dagger conveniently to hand,
with which he struck back, causing Choronzon to retreat. The spirit
had one last go at trying to seduce Neuberg, and then gave up completely
and disappeared. [156]

The temptation is to dismiss all this out of hand; but if we reject it
as nothing but juvenile fun and games, we are thumbing our nose at
thousands of years of penetrating study. Have we the right to do so?

One of the very few contemporary magicians to write intelligibly
on the subject, J. H. Brennan, [12] has described being present at an attempt
to conjure up a spirit. A Hebrew ritual from the Golden Dawn's
collection was used. Besides Brennan in the magic circle was an assistant,
and an attractive young lady named Miranda. Amid clouds of swirling
incense the three of them spent two and a half hours in the confined
space of the circle in non-stop invocation. Brennan describes the climax:

> All these factors must have a cumulative effect, which is why I must end
> the story on a note of bathos. I am not sure if the evocation was brought
> to a successful conclusion or not.
>
> It was an elemental evocation. That is to say, the force being called up
> was neither divine nor infernal, but neutral. At the climax of the ritual I stared
> with tear-filled eyes towards the triangle, half-hidden in the fog of billowing
> smoke. Was there a shape there? One moment my answer was Yes: the next
> I was sure it was No. The smoke seemed to form a slimly-built, narrow-
> faced man . . . then the illusion vanished — if it was an illusion.

As Brennan himself admits, in goings-on like this the magician works himself up into quite a state. Over and above his own enthusiasm, sharpened by whatever motivations may be spurring him in his enterprise, he makes use of many devices to heighten the sense of other-worldliness — a highly ornamented setting, heavily invested with symbolic paraphernalia, is *de rigueur*, as are fine robes, ritual daggers and regalia. The words employed have been formulated to roll sonorously off the tongue and to bring to the minds of all those present the superhuman powers and authorities they are consorting with. At the same time they are likely to have in their minds the sense of a continuing tradition, of a long line of hierophants going back to ancient Egypt or Chaldea, of time-honoured rituals whose origins are lost in the mists of antiquity and which have been hallowed and strengthened by continual use across the centuries . . . And then there are the details of the actual ceremonial selected for the occasion: the use of drugs, the sacrifice of animals, the heterosexual, homosexual or masturbatory practices up to and including a carefully timed orgasm — all these taken together will certainly put the practitioners into a very unusual state of mind, and very likely into an altered state of consciousness.

During that altered state, it is not unlikely that hallucinations will occur; and so the belief that the ritual has been successful, and an entity really summoned, may be no more than a hallucination, an illusion created by the state of the magician's mind.

But the possibility remains that it could be something more. There are indications that in certain altered states, some people are occasionally capable of physical feats — telekinesis, healing, out-of-the-body journeys; and as we have seen, it does seem as though this can result in some kind of projection that is sufficiently substantial to be seen by others. So we cannot rule out the possibility that this is what may happen as the result of a magical practice: that what Neuberg saw out there in the Algerian desert was not a hallucination created by his own mind, nor Crowley himself playing silly games, but some kind of entity emanating from either his own mind or Crowley's, possessing a real if ephemeral existence.

This is a possibility that author Richard Cavendish is ready to accept, explaining it in these terms:

Although such manifestations may often be the results of hallucination, or sometimes of deliberate deception, occultists believe that this is not always the case. They say that self-intoxicating procedures are necessary because the spirit is not a part of the normal, everyday world, and so cannot be experienced in normal states of mind. The spirit may show itself in a form created for it by the magician's imagination, but it is a real force. It may be a force or intelligence that exists independently of the magician, and if so it is no more imaginary than the forces of electricity or gravity, or it may come from within the magician himself, in which case it is no less real than the forces of ambition or pride or desire which we recognize in ourselves. Behind all this, of course, is the perplexing question of what reality consists of, and whether it is sensible to regard those things which we seem to perceive in 'normal' conditions as real, but those which we seem to perceive in 'abnormal' states of mind as unreal.[16]

Which is, of course, the central dilemma of this entire study. We have seen that other experiences, undergone in a very different spirit from that of Crowley and Neuberg, can result in something that seems to be something more than a simple hallucination: we cannot deny, in their case also, at least the possibility that something more substantial was conjured forth.

2.6   THOUGHT-FORMS

One of the many awkward questions you can put to a believer in non-human entities is: What are they doing when they are not plaguing — or protecting — their human percipients? When we considered the case of the lady climber rescued by the oriental gentleman, we noted the improbability that he hung around that mountain permanently in case he should be needed. So, where had he been, and what had he been doing, that he could stop doing whatever it was and get so quickly to where he was needed?

I am deliberately being naive in putting the dilemma in such terms, but that is because I have never found explanations put in more high-sounding terms to be satisfactory. For if these entities have a physical existence, then it is an *on-going* existence. I will allow that, between jobs, they may be utterly dormant, like seeds in the Australian desert waiting for the once-in-five-years rain that will bring them to life: but even dormancy is an event — or perhaps a non-event — *in time.*

Those who claim to know about these things (let us not question the source of their knowledge but give them for the moment the benefit of our doubts) assert that such objections are beside the point, because for these entities, time does not have the meaning it does for us: we alone of God's creatures, it seems, are time-bound. What we have done to deserve this imprisonment is something the theologians are only too ready to explain, but it seems to me that this reckless surmise has no other justification than that it gets them out of a logical difficulty. 'Ten thousand ages in Thy sight are like an evening gone' was all very well for nineteenth century notions of the universe, but after Einstein had taken it in hand and redesigned its topography, it became necessary for supernal beings to be lifted out of time altogether.

When they visit us, however, or come hurrying in answer to our summons, they enter our time dimension, and subject themselves to its restraints, though these are frequently over-ridden. The fundamental paradox here is that human affairs are meaningless without the concept of time, and would therefore seem meaningless to any entity who had no conception of time. A Boddhisattva could not take seriously the plight of a distressed climber, a Mephistopheles could not respond to Faust's command, or the Virgin Mary come on a slum-visiting mission to underprivileged Catholic teenagers, without an understanding of the concept of time and a belief in the importance of things that happen in time.

To be aware of our timeboundedness, they would have to be aware of their own timelessness; but there can be no timelessness unless time itself exists. To exist outside time, there has to be a time to exist outside of; and however much you play with notions of folds in time, the fact remains that the Virgin can pay a succession of visits to Bernadette Soubirous only if she has some kind of calendar to refer to.

So it is very much easier to believe in entities that exist within our own time continuum and never leave it. Instead, the problem of what they are doing between visitations is resolved in a different way: they cease to exist. They come into being only when they are needed.

Of such are the entities created by our projection experimenters, who seem able to create or call up some kind of 'double' of themselves, which thereupon proceeds to act with apparent autonomy during its span of existence. But that span of existence is determined, generally unconsciously, by its human creator.

The *thought-form* represents an alternative model. Like the projected double, it exists within its creator's own time scale: but it differs from the double by having not only an autonomous freedom of action, but also a persona of its own.

We have already looked at one category of entities that seem as though they might be an example of this — a category which, indeed, might almost as fittingly have been placed in Part Two as in Part One of this inquiry: the imaginary companions created by children, and more rarely by adults. I use the word 'created' without wishing to imply a conscious process; all the indications are that it is more usually unconscious. However, there are some instances of a deliberate act of creation, such as the following.

George Estabrooks, a psychologist and hypnotist, was once recuperating in a hospital. To pass the time, he hypnotized himself so that he could enjoy a vision of a pet polar bear, which 'I was able to call up merely by counting to five. This animal would parade around the hospital ward in most convincing fashion, over and under the beds, kiss the nurses and bite the doctors. It was very curious to note how obedient he was to "mental" commands, even jumping out of a three-storey window on demand.'

Estabrooks' creation was never anything but fantasy; its creator knew that everything about it, including its apparent autonomy, was illusory. But how different was this creature, casually devised as a pastime, from a seriously created thought-form entity such as the following, the best-known case in the literature?

Alexandra David-Neel was a redoubtable French traveller and author: she was also a practising Buddhist who spent fourteen years in Tibet studying the religion, and in particular its occult practices, some of which she sought to practise herself:

Must we credit these strange accounts of rebellious 'materializations', phantoms which have become real beings, or must we reject them all as mere fantastic tales and wild products of imagination? . . . Allowing for a great deal of exaggeration and sensational addition, I could hardly deny the possibility of visualizing and animating a *tulpa*. Besides having had few opportunities of seeing thought-forms, my habitual incredulity led me to make experiments for myself, and my efforts were attended with some success. In order to avoid being influenced by the forms of the lamaist deities, which I saw daily round me in paintings and images, I chose for my experiment a most insignificant character: a monk, short and fat, of an innocent and jolly type.

I shut myself up in *tsams* (states of meditation in seclusion) and proceeded to perform the prescribed concentration of thought and other rites. After a few months the phantom monk was formed. His form grew gradually *fixed* and life-like looking. He became a kind of guest, living in my apartment. I then broke my seclusion and started for a tour, with my servants and tents.

The monk included himself in the party. Though I lived in the open, riding on horseback for miles each day, the illusion persisted. I saw the fat *trapa* (student monk), now and then it was not necessary for me to think of him to make him appear. The phantom performed various actions of the kind that are natural to travellers and that I had not commanded. For instance, he walked, stopped, looked around him. The illusion was mostly visual, but sometimes I felt as if a robe was lightly rubbing against me and once a hand seemed to touch my shoulder.

The features which I had imagined, when building my phantom, gradually underwent a change. The fat, chubby-cheeked fellow grew leaner, his face assumed a vaguely mocking, sly, malignant look. He became troublesome and bold. In brief, he escaped my control.

Once, a herdsman who brought me a present of butter saw the *tulpa* in my tent and took it for a live lama.

I ought to have let the phenomenon follow its course, but the presence of that unwanted companion began to prove trying to my nerves; it turned into a 'day-nightmare' . . . I decided to dissolve the phantom. I succeeded, but only after six months of hard struggle. My mind-creature was tenacious to life.

There is nothing strange in the fact that I may have created my own hallucination. The interesting point is that in these cases of materialization, others see the thought-forms that have been created.

Tibetans disagree in their explanations of such phenomena: some think a material form is really brought into being, others consider the apparition as a mere case of suggestion, the creator's thought impressing others and causing them to see what he himself sees.[28]

Clearly, from an evidential point of view, Alexandra David-Neel's account is scientifically worthless: it depends wholly on her own unconfirmed and uncorroborated statement, and she is not able to supply a scrap of supportive evidence. While on the whole the author maintains a fresh, somewhat sceptical attitude towards the business, she is not entirely naive: this episode, the most dramatic in her book, is placed at its very end as a calculated climax, which implies a deliberate intention of exciting the reader, which seems somewhat out of keeping with the high moral tone of her spiritual claims. This must make us wonder how much more may have been included, consciously or otherwise, with the same intention?

What is more disturbing is that the case is virtually unique. Her book was immensely popular, and has remained a classic; it is hard to believe that among all her tens or even hundreds of thousands of readers, there has not been one who has sought to duplicate her feat. But if so, we have not heard of any successful replication. Was Alexandra David-Neel specially gifted? She claimed no more than that she was doing something that was standard, if infrequent, practice among the Buddhists of Tibet; perhaps among them there may have been numerous cases of which we have not been told. Yet if so, given the intense interest we westerners have taken in oriental religions during the past hundred years, it is a surprising omission.

Significantly, one of the few comparable cases on record also emanates from the same source of inspiration.

Nicholas Mamontoff tells how, in December 1912, a group of Russian members of an occult-scientific investigation group, called the Brotherhood of the Rising Sun, used to meet in St Petersburg. On one particular occasion they invited a guru — a teacher — whose appearance was Chinese or Tibetan, though he spoke Russian, to instruct them in his knowledge. He led them in an experiment to create an *egrigor*, an entity that seems to have been very similar to Alexandra David-Neel's *tulpa*:

> *Egrigors* are created by human thoughts. As we know, our thoughts consist of electric energy plus vital fluid or *pranah*, possessed by every living body. A thought, having issued from our mind, produces some kind of reflection in the astralic plane and stays there for a certain period. When it is not 'fed' or supported by similar emotions or thoughts issued by us or by other persons, such a reflection vanishes without a trace.
>
> The ordinary thoughts we produce from moment to moment do not make *egrigors* ... An accumulation or concentration of evil or good thoughts creates *egrigors* so real that we can perceive them even by our imperfect organs of sense.
>
> Especially powerful and dangerous emanations are produced by human beings dying violent deaths. Cursing one's murderer or executioner creates a dreadful *egrigor*, which lives in the astralic plane for a long time. Having received, from the healthy, strong body of the murdered or executed person, a lot of *pranah*, such an *egrigor* becomes so vital that it can cause trouble even in our physical plane. The vitality of *egrigors* varies. Some are motionless with unclear and diffused forms. They look like embryos and are the weakest type of *egrigor*. More powerful thoughts create more vital *egrigors*.

After the guru had explained the principles, the Grand Duke, in whose house the meeting was being held, invited the guests into a room from which all furniture, except chairs, had been removed. The lights were turned off. The guru asked all present to concentrate their thoughts upon something unusual, something that had never existed in reality. Somebody proposed a dragon.

'Why do you choose such an ugly thing?' asked the guru. 'We already have a lot of evil things in the astralic plane. Let us concentrate on something harmless and funny — on the Cat-In-Boots, for instance.'

In order to make the *egrigor* clear the guru gave a description of the cat. He should be red-haired, dressed as a musketeer in a blue cloak, red coat, and high boots with spurs.

More than half an hour of concentrated thinking passed without any result and the group began to doubt that their experiment would succeed. 'Concentrate! Concentrate! Do not think about anything but the cat,' urged the guru in the darkness.

More minutes passed and then, in the absolute darkness, a light cloud appeared. It moved and changed shape until finally it assumed the unclear form of a cat. A few minutes later the details of the animal's body became clear but its dress remained only multi-coloured vibrations.

'Obviously you cannot imagine the musketeer's uniform,' said the guru, 'Your thoughts are various. Now, think only of a red-haired cat wearing common Russian boots.'

Within a few minutes the features of the cat stabilized and on his hind feet was a pair of Russian boots. The *egrigor* was motionless and looked like a poorly developed photograph. 'Do not think about the cat any more and watch what happens to it,' ordered the guru. Sitting in the darkness the audience saw the form of the cat gradually melt and at last disappear completely.

'The duration of an *egrigor*'s life and its vitality are in direct proportion to the tension, energy and pranah spent in its creation by the human mind,' explained the guru.

After the seance my father asked the guru whether it is possible for a single person to create an *egrigor*. He replied that it is very difficult and even impossible for the average person. However, with special training and exercises it can be done, he said. 'The Western scientists never realize how powerful the human mind is and what miracles it can work.'[42d]

Two cases, neither of them of unassailable authority, and neither confirmed by any supportive evidence, are hardly enough to convince us that such feats are possible: at the same time, we cannot reject the testimony out of hand when we have no reason to suppose that it is given in anything but good faith. The circumstances under which these thought-forms were created have a certain logic, and the resemblance to other cases in our study is evident. So, albeit with reservations, we should be prepared to include them in our inquiry.

There is one aspect of Alexandra David-Neel's experience we should note if we are to be impartial in our search for explanations. She tells us how her *tulpa* gradually changed its nature; from fat and jolly it transformed itself into something sly and malignant. This was surely not what its creator had in mind — not at any rate in *conscious* mind; and while a psychoanalyst might well come up with a plausible reason why this product of her subconscious mind differed from what her conscious mind intended, and came to represent, let us suppose, her guilt or fear in dabbling in such occult practices, there are alternative explanations. Some Christians, for instance, would say that in creating this fantasy-creature, she provided an 'empty room', swept bare, into which a demon could and did step and take up its lodging. That is to say, her innocent puppet was possessed by a real and perhaps evil spirit, of the same kind that we looked at in Part One, or perhaps simply an elemental such as magicians like Crowley have sought to summon up. It is yet another possibility that we do not have the right to ignore.

In the 1970s an experiment was carried out which, though it was not successful in producing a visible entity, was sufficiently related to the experimental processes we have been looking at to justify our attention. The experiment was performed, over a period of more than a year, by a group of people in Toronto, under the leadership of the eminent psychical researcher A. R. G. Owen. They managed to create a being named Philip, for whom they had previously devised a wholly fictitious identity: by an act of collective will not unlike that of the St Petersburg circle, they succeeded in getting Philip to communicate with them at sessions conducted more or less like spiritualist seances. Though there was never any serious doubt in his creators' minds that he was simply a creature of their imaginations, they were astonished at the degree of autonomy he displayed, giving them messages and details of his life when supposedly living on earth which seemed quite independent of the conscious minds of the group. [114]

The implication of the Philip experiment for the whole question of mediumistic evidence for survival after death is, of course, profound: inevitably it casts doubt on a great many supposed communications with spirits of the dead. Certainly it bears out the Russian guru's comment on the unrealized powers of the mind.

Naturally, efforts were made to see if the experiment could be carried a stage further, to get Philip to manifest in visible form. Unfortunately these attempts failed: he was never more than a voice, though he may also be credited with a number of psychokinetic feats that occurred at the meetings of the circle. Yet, even though this failure means that Philip is not directly relevant to our inquiry, the fact that an imaginary being can be conjured up experimentally, even to that degree, gives some support to the accounts of Alexandra David-Neel and Nicholas Mamontoff.

It is ironic that a team of psychical researchers should fail to create a visible entity, for if the occultist Dion Fortune can be believed, it is sometimes possible for a psychically gifted person to see a thought-form created by someone even though its creator is the only other one to see it:

There are certain types of insanity in which the lunatic believes himself to be the victim of an attack by invisible beings, who threaten and abuse him and offer base or dangerous insinuations. He will describe his tormentors, or point to their position in the room. A psychic who investigates such a case can very often see the alleged entities just where the lunatic says they are. Nevertheless, the psychologist can come forward and prove beyond any reasonable doubt that the so-called 'hallucinations' are due to repressed instincts giving rise to dissociated complexes of ideas in the patient's own subconscious mind. Does this mean that the psychic is mistaken in thinking he perceives an astral entity? In my opinion both psychic and psychologist are right, and their findings are mutually explanatory. What the psychic sees is the dissociated complex extruded from the aura as a thought-form. A great deal of relief can be given to lunatics by breaking up the thought-forms that are surrounding them, but unfortunately the relief is short-lived; for unless the cause of the illness can be dealt with, a fresh batch of thought-forms is built up as soon as the original ones are destroyed. [45]

Whether or not we accept Dion Fortune's terminology, which is derived from a very specific belief-system, in broad terms the process she describes is logical and provides us with, at the least, a useful working hypothesis. Thought-forms of any kind seem to result from intense mental activity, conscious or unconscious, and when it is unconscious — as, except in the David-Neel and Mamontoff cases, it generally seems to be — it is likely to be stimulated by stress which is usually unpleasant. But this is not necessarily the case. In the act of artistic — usually literary — creation, many creators of fictional characters have spoken of them as being more living than their fellow-humans, and even seeming to impose a sense of presence that can amount to a full-blown visual hallucination. What is more, there is at least one case in which a psychically gifted person saw an author accompanied by a lady she did not recognize: when she described the lady to the author, he acknowledged that it was a character from his current book, in which he was deeply involved at the time.

2.7 SPIRITS OF THE DEAD?

Since the middle of the nineteenth century, many people have come to believe that living people can communicate with former inhabitants of this world who have subsequently passed on to another plane of existence.

This belief has been embodied in an institutionalized belief-system, Spiritualism, which often takes on the character of a religion: its adherents number millions throughout the world. It has accumulated an enormous quantity of evidence, some of which is indeed most easily interpreted in terms of Spiritualist belief. This is not to say, however, that it has been definitely proved to be so; simply that alternative explanations for some of the reported phenomena are so devious and cumbersome that they carry little conviction except to those determined at all costs to find an alternative hypothesis that does not presume survival after death.

None the less, after more than a century, a hypothesis is all that the Spiritualists' beliefs remain. The conditions under which evidence has been obtained, the inconsistency of the phenomena, the undoubted prevalence of conscious fraud on a very wide scale, and an incalculable amount of unconscious fraud and self-deception besides — these and many other factors work against the ready acceptance of the Spiritualist hypothesis. And there are many who, while accepting that the reported phenomena are often genuine, see no necessity for supposing that they necessarily require the Spiritualist interpretation. The fact that a spirit medium who produces psychic effects gives the credit to her spirit controls should, no doubt, be listened to with respect; but it should no more be accepted without question than should Bernadette Soubirous' assertion that it was the Virgin Mary who appeared to her at Lourdes.

The alternative hypothesis must also be taken into account; which is that the admittedly genuine phenomena of Spiritualism can be accounted for by explanations that do not presuppose the survival and communication of the dead. What these alternative explanations may be, we shall not find it easy to say: but some fruitful suggestions arise from the comparative study we are currently undertaking. We shall find that, while we cannot reject outright the Spiritualist hypothesis, equally we cannot discount the possibility that similar phenomena could result from psychological and cultural forces manifesting via certain processes.

One force must be the natural human desire to obtain proof that there is more to existence than our life on Earth; proof that we survive, in whatever form, would be immensely comforting to most of us, though a minority profess to be alarmed at the prospect. From the start, the main burden on the Spiritualists has been to furnish proof that will convince the public that they will survive death: the history of the Spiritualist movement has consisted in a steadily escalating development in the weight of proof offered. While explicable in sociological terms, this escalation does somewhat support the sceptic's assertion that the whole business is an artefact, constructed by mankind for its own comfort.

The process began, for all practical purposes, in 1848, when in the state of New York the Fox sisters discovered that unknown forces were responding to raps made by themselves. By employing a code, these rapped responses became verbal messages; this led to a search for a more fluent method of communication, and table turning, glass moving, planchette and automatic writing represent increasingly convenient techniques. Then came a bigger step, to verbal communication, via a medium who claimed to be able to relay messages from the dead: this in turn led to direct voice communication, in which the medium purported to be actually taken over by the dead person, or by a spirit control acting on the dead person's behalf.

At the same time the physical manifestations, which had begun with the rapping sounds, continued, for they provided confirming evidence that the dead have the power to do things that are not possible by earthly laws, and which are therefore convincing signs of their existence and of the validity of their claims. Thus came about the whole range of seance-room physical phenomena — the levitated trumpets and other articles, the tambourines played in mid-air by unseen hands, the levitation of the medium and of furniture, slate writing, and so on. Many people who might have been inclined to give a serious hearing to the Spiritualists were disgusted by these performances, and it is arguable that they have done more harm than good to the movement — that evidence of this kind is worse than none at all.

In one respect they certainly did harm, in that they inextricably confused the notion of superhuman powers with the Spiritualist hypothesis; so that whenever a medium performed or caused to perform something that was beyond normal human capability, it was liable to be interpreted as a sign or a gift conferred by the spirits. This is a wholly unwarrantable assumption, but one that has resulted in a widespread fallacy, largely held even today: that there is a necessary causal connection between psychical phenomena and the Spiritualist hypothesis.

However, in the case of the most impressive physical evidence of all, that causal connection was indeed inherent; this was the phenomenon of materialization, the actual manifestation of the spirit of the dead person, the logical culmination of the search for ever better evidence. However, the escalating development of evidence can be seen in two different ways. If you believe that the mediums are indeed in communication with the surviving dead, and that those dead are indeed seeking ways of providing the living with evidence for survival, then it is logical to suppose that they would gradually improve their methods of communication, elaborating ever better techniques.

If, on the other hand, you fear the Spiritualists are deluding themselves, or are being deluded by false spirits, then you may feel that the progressively more sensational displays are more of the nature of circus acts, in which each performer seeks to outshine his predecessors in order to capture the attention of the public.

Unfortunately for our attempts to evaluate these matters objectively, the term 'circus' is a very apt one. Though some mediums have been dedicated people who have worked hard and given their services freely in the cause of their belief, others have been more professional in their attitude, making a living in proportion to their ability to furnish the proof their clients desired; so there was a pressure on them, as well as a vested interest, to produce results. The more impressive those results, the more the mediums themselves would benefit both in prestige and in more tangible ways. Consequently we are faced with the paradox that the better the evidence, the stronger are our grounds for suspecting it.

A parallel corollary is that not only is the incentive to deceit very strong as regards the medium's own interests, but it is abetted by the public, who have displayed in astonishing ways a willingness to be deceived. The history of Spiritualism displays humanity at its most credulous: even after a medium has been unequivocally exposed as fraudulent, his faithful followers continue to believe in him and rationalize their trust by insisting that the exposure was itself a fraud, dictated by spite or malice; that the medium was himself deceived by malicious forces; or, in extreme cases where the evidence of fraud cannot be dodged, that an otherwise genuine medium has temporarily resorted to fraud with the best intentions, reluctant to disappoint his clients, or even that he had no say in the matter but was deliberately driven to cheat by diabolical forces!

It is no part of this inquiry to get any deeper into the squalid disputes attached to Spiritualism than is necessary, but it is important that the cultural context in which our phenomena occur should be appreciated. Ultimately, however, only two points concern us here:

★ Do genuine cases of materialization of alleged spirits of the dead occur?

★ If they do, are they really spirits of the dead or not?

Typical of the earliest materialization demonstrations were those
presented by the Eddy family of Vermont, USA, in the early 1870s.
The Eddys were a farming family, but put on a show almost every
evening for the benefit of clients who came and lodged at the house,
paying, it must be admitted, only a modest fee to cover their lodging.
The shows consisted of William Eddy sitting in a cabinet in a dark room
and causing spirits to be materialized. The audience were required to
sing, music was played, and the sitters in the front row were asked to
join hands. A person who was present at the Eddys' demonstrations
gave this account:

> The only persons allowed to take a seat upon the platform were a Mr Pritchard
> and a Mrs Cleveland; dear gullible souls who could be readily psychologized
> into believing that they were eating a piece of the moon in shape of green
> cheese. These both touched and conversed with the substantial shadows which
> stepped cautiously from the door of the cabinet, as if making sure that some
> investigator were not ready to spring upon them; and occasionally went
> through the shuffling manoeuvres characterized as dancing, while not one
> of the audience was permitted to advance near enough to distinguish their
> features in the distressingly subdued light of the solitary lamp.
>   Up we went, and sat on the benches prepared for us. All on the front row
> joined hands 'to make a magnetic chain' though where the magnetic current
> went at either end, I don't know. The man with the violin played vigorously.
> The light, back of us all, was turned so low it was impossible to distinguish
> features two feet from us. When the curtain over the door of the cabinet
> was lifted, and a 'form' appeared, the violin was hushed; the 'form' was mum,
> until the front bench, beginning with no 1, commenced to ask the stereotyped
> question, 'For me, For me,' until the form bowed. 'Is it father?' — no response.
> 'Is it uncle?' — still silent. 'Is it Charlie Myrtle?' — the form bows. 'Why,
> Charlie, I'm so glad to see you!' Then it is reported that 'Charlie Myrtle' was
> distinctly recognized, and that is about all there was of it . . . We could detect
> no fraud, as all real opportunity for investigation had been dexterously cut
> off; but that we were in the presence of real spirits we never for once believed. [72]

And yet hundreds of people came, and went away satisfied after seeing
and speaking with dead relatives and friends; on one evening eighteen
figures of different ages were seen, sometimes two or three at a time:
the theosophist Olcott, visiting in 1874 and staying for several months,
saw several hundred different entities in the course of his stay. [71] Here
is another witness from the same epoch, a Dr Horace Furness:

> A woman, a visitor, led from the cabinet to me a materialized spirit, whom
> she introduced to me as 'her daughter, her dear, darling daughter', while
> nothing could be clearer to me than the features of the medium in every line
> and lineament. Again and again men have led round the circle the materialized
> spirits of their wives, and introduced them to each visitor in turn; fathers
> have taken round their daughters, and I have seen widows sob in the arms

of their dead husbands. Testimony such as this staggers me. Have I been smitten with colour-blindness? Before me, as far as I can detect, stands the very medium herself, in shape, size, form and feature true to a line; and yet, one after another, honest men and women at my side, within ten minutes of each other, assert that she is the absolute counterpart of their nearest and dearest friend, nay, that she *is* that friend.[119]

The wonder is that anyone, confronted with evidence like that, took the claims of the Spiritualists seriously. But there were other cases which were not so easily put aside; despite the absurdities, despite their own scepticism, objective investigators had to confess that there was a case to be answered:

I shall not waste time in stating the absurdities, almost the impossibilities, from a psycho-physiological point of view, of this phenomenon. A living being, or living matter, formed under our eyes, which has its proper warmth, apparently a circulation of blood, and a physiological respiration, which has also a kind of psychic personality, having a will distinct from the will of the medium, in a word, a new human being! This is surely the climax of marvels! *Nevertheless it is a fact!*[130]

That was the eminent French scientist Charles Richet, commenting on the materialization phenomena he had personally observed and investigated over a period of years. With the greatest reluctance, which it is hard to believe he only pretended to feel, he had come to the belief that materialization is a fact, and that there is no possible way of disposing of that fact by attributing it to fraud or faulty observation or misinterpretation. The fact itself had to be accepted.

This is not to say that the Spiritualist interpretation had also to be accepted. Richet himself was one of the few with the perception to recognize that it was by no means a necessary consequence. In his view, the Spiritualists were accounting for the phenomena as best they could, in terms of their own belief-system.

Although materialization phenomena have been reported intermittently over a period of about a hundred years, and are still occasionally reported today, they have only been scientifically investigated for a small part of that time, during a period roughly between 1900 and 1930. Before 1900, though many instances were reported, little attempt at scientific investigation was carried out: after 1930, for whatever reason, the manifestations gradually died out almost entirely. It is tempting to attribute this to the discouragement caused by hard-headed investigation, which exposed so many fraudulent cases, but this is hardly consistent with the fact that many of those respected investigators who had devoted most time to the matter became convinced that some at least of the phenomena were genuine.

It seems more probable that it was not so much that the manifestations died out of their own accord, as that they ceased to be produced because of a switch of interest. The main thrust of scientific psychical research was now directed towards laboratory testing of extra-sensory powers, pioneered by Rhine and others, all of which seemed more in keeping with modern science; seance room phenomena, by comparison, seemed fit only for relegation to the same limbo as that into which phrenology, astrology and mesmerism had been consigned. The implication is, I know, that these phenomena occur as a consequence of demand: when interest and incentive are absent, the phenomena themselves fail to occur. I make this implication deliberately, because I believe it is a significant clue to the true nature of the phenomena. But I must make it clear that it in no way implies that the phenomena are any the less genuine.

Here is one of the cases that convinced Richet. It occurred while he was investigating Marthe Béraud, at his home at the Villa Carmen on an island off the south coast of France, in 1904:

> I saw a fully organized form rise from the floor. At first it was only a white, opaque spot like a handkerchief lying on the ground before the curtain, then this handkerchief quickly assumed the form of a human head, level with the floor, and a few moments later it rose up in a straight line and became a small man enveloped in a kind of white burnous, who took two or three halting steps in front of the curtain and then sank to the floor and disappeared as if through a trap-door. But there was no trap-door. [130]

This case is typical of perhaps the majority, in which the phenomenon occurs as the result of an action deliberately induced by the medium: that is, the entity appears as a consequence of an act of will on the medium's part, the medium herself being in an altered state of consciousness — a trance. In this respect, the phenomenon resembles the deliberately contrived 'astral visits' we have already considered. Another feature is that the medium often emphasizes the need for supportive power from the other sitters: whether this is true or not, it is a feature that we noted in the case of the thought-form conjured up on the guru's instructions in the Russian circle. So there are, ostensibly, features in common between mediumistic materializations and other forms of voluntary entity-experience.

The possibility of fraud is something that no serious investigator has ever discounted. Even in the 'primitive' era of materialization, stringent precautions against deception were taken, a team of people searching the room and the medium herself, watching closely throughout the proceedings: nineteenth-century investigators were no more happy about being gulled than we are today. Often the medium was on her own, without a confederate, surrounded by a group of people who, whether or not they were sympathetic, were alive to every possibility of cheating and watching carefully to detect the least sign. Only when they had control of the conditions — as was decidedly *not* the case at the Eddy seances — and only when the medium submitted to their requests, were they prepared to admit that the phenomena were genuine.

Consider what would be necessary to produce by fraud the kind of phenomena observed by Richet:

★ the medium would have to conceal on his person the raw materials for the construction of the entity, in such a way as to escape the detection of a group of determined searchers.

★ the medium would then, in the course of giving a performance in which he was the focus of attention, have to produce those materials from their hiding place, and fashion them into a figure whose movements could appear to be autonomous, and whose appearance was sufficiently lifelike to persuade the vigilant observers that a human form was present, warm and lifelike to the touch.

★ the substance of the entity, though visually solid and tangible, would have to be of such a kind that it could be felt by the investigators to gradually melt away to nothing while being held by them — or at any rate, to give the illusion of so doing.

★ the 'entity' would then have to be returned to its former constituents, and be concealed once again in the room or on the medium's person, in such a way as to escape further search.

★ this performance would have to be repeated time and time again, in the presence of different investigators, who were capable of devising elaborate tests, such as making the medium swallow a substance that would colour anything he had in his mouth, throat or stomach; he would have to pass such tests, sometimes imposed on the spur of the moment, without incurring suspicion by evasion or refusal.

Yet it was under such conditions that mediums were able to produce phenomena that could provoke this kind of account from an investigator:

First, a shapeless mass is seen; this may either emanate from some portion of the medium's body, either the chest or head, or very often the mouth, or it may be independent of the medium. It is generally, but not always, faintly luminous, and moves apparently of its own accord. The mass is gelatinous and cold. It gradually takes shape and gains in consistency, until it reproduces, to the touch, every characteristic feature of a normal human hand. In the very great majority of cases, it is a hand, or a hand and arm, that appear, seldom or never feet or legs. Heads, faces, and complete human forms have been seen in certain cases, but rarely; they have then often been characterized by the fact that they appear to be flat, devoid of any relief. [113]

Similar cases could be cited from half a dozen investigators of the period — Maxwell, Gibier, Sudre, Schrenck-Notzing, Geley and others. Marthe Béraud, investigated by Schrenck-Notzing and Mme Bisson as well as by Richet, was observed and controlled by some two hundred people at various times, of whom a good many were highly sceptical: yet she was never detected in any fraud.

We may therefore take it that the evidence for materialization is as good as, if not better than, that for almost any other kind of entity we are discussing in this inquiry. At the same time, we must recognize that *what* is happening is different in many ways. Not only is this a voluntary experiment, but it is carried out *as* an experiment; the medium is operating under clinical conditions, in the presence of sceptics. It is a cold, calculating effort to secure proof for science, not comfort for unhappy bereaved persons. In this unemotional climate, any question that the purpose behind it all is to set up a channel of communication between this world and the next has been either forgotten or shelved. What is astonishing is that the mediums put up with it, as such people as Marthe Béraud often did for years on end.

Only rarely did the resulting entities resemble people who could be identified. More often they were not identifiable; frequently they were not even lifelike. The best-known of the physical mediums, Eusapia Palladino, only rarely produced complete materializations: these were 'a very large man' who emitted red light and gave every appearance of life, the figure of a child who resembled a child known to the investigators, and two unidentified girls 'of an oriental type'. [155]

Unsatisfactory in a different way were the figures produced by Marthe Béraud for Schrenck-Notzing. [9, 137] She worked for him and his colleagues for four years, during which time she gradually developed from producing mere fingers and hands to more complete figures. But while the hands were three-dimensional and lifelike, the complete figures were two-dimensional and very far from lifelike. What is more, there were even glaringly obvious anomalies, such as paper folds and printed lettering, plainly visible.

Inevitably this led to accusations of fraud; somehow, it seemed obvious, the medium had managed to smuggle in cut-out pictures, concealed about her person, and build them up into figures. The fact that the faces could often be identified as actresses, politicians and so on was indicative of their source, and when the letters 'LE MIRO' — evidently part of 'LE MIROIR' — were detected, this seemed to clinch it.

Except that it was done so blatantly — and so badly. Surely even the most stupid hoaxer would have cut off the give-away lettering from beside the picture? Besides, on further examination it was found that though the faces in Marthe's materializations were *like* those in the magazines, they were not *identical*. Efforts were made to reproduce the alleged fake; they failed completely. The conclusion must be that where the investigators themselves, who had no cause to conceal their movements, could not manage the trick, it is next to impossible to suppose that a single girl, who had been searched beforehand, could, under the vigilant eyes of more than one investigator, carry out the trick undetected, leaving no trace afterwards.

But why, if she was producing a genuine materialization, did she produce such ridiculously unlifelike figures? We do not have an explanation, but we have a comparison that may be useful even though it is with a phenomenon hardly less controversial. The 'thoughtographs' of Ted Serios, [34] consisting of images somehow imprinted by his thoughts on to virgin photographic film, have something of the same half-real quality that is displayed by Marthe's figures. If she was seeking to reproduce a face, just as Ted was seeking to reproduce a scene, the fact that they produced somewhat similar results suggests that they used somewhat the same process — for instance, that their subconscious minds went around foraging for suitable material, derived it from magazine photographs seen but not remembered consciously yet stored somehow in the subconscious, and imprinted it — not very well — on photographic film in Ted's case, on ectoplasm in Marthe's. (Let me hastily emphasize that I am simply trying to propose a method, drawing on our knowledge of what others have achieved, of how it might have been done. This, or something like it, must be the *kind* of explanation which we have to be looking for, because it is the only kind of explanation which will meet the facts.)

One of the alleged facts, of course, is the existence of the building material which seems to be used in the process of materialization. The very existence of ectoplasm has been frequently questioned, and undoubtedly in some cases it has been shown to be mucus, in others cheesecloth. It has been seen by thousands, and felt by a good many; it has frequently been photographed, but it has never convincingly been retained and analyzed. None the less, there do seem to be quite good grounds for believing that there exists a substance which certain people, in a certain state, are able to extract or excrete from their bodies and form more or less successfully into whatever shape they choose, and which, when in that shape, is capable of being to some degree animated, possibly even given temporarily a capacity for autonomous motion, which may or may not be under the control of the medium's subconscious mind; or alternatively, if you prefer, under the control of the spirits of the dead who have for the moment taken over the medium's mind.

Ectoplasm would seem to be essentially a part of the medium's body, for when it is weighed it has substance, and its weight has been shown to correspond to that simultaneously lost by the medium. It emerges generally from one or more of the natural orifices of the body — the mouth, the nipples, the genitals; but also from the fingertips and elsewhere. It first emerges in a more or less shapeless state, often as a mist which may be luminous; the shape is acquired gradually. It can pass through loosely woven nets and veils, but not, it seems, through more tightly woven clothing.

The existence of such a substance is, of course, totally contrary to all that we know of the human body; there is not a scrap of biological evidence for it. Considering how much we know of our physical organism, this would seem to be sufficient to disprove it altogether; but then how are we to explain the evidence? Crawford, measuring the weight loss of Kate Goligher, found that she would normally lose 7kg out of her total of 62kg when emitting ectoplasm, and that, with great effort, she could shed as much as 24kg — that is to say, well over a third of her body weight. At such times, the weight of the ectoplasmic structures was exactly that which the body had lost. The only conceivable explanation is that a proportion of the medium's body had been turned into ectoplasm; a process totally contrary to everything we know of physiology. No wonder most people prefer to deny the facts! [23]

In order to obtain these results, Crawford had — through Kate — to obtain permission from the spirits who allegedly communicated through her; they allowed it only after repeated requests, and then only under specified conditions. None of those involved seem to have had any doubt that this was a clear case of spirit activity; we, however, are under no obligation to accept this at face value. If we prefer, we may surmise that the permission was accorded by Kate's own subconscious mind, which for psychological reasons she preferred to exteriorize as 'spirits'.

Whatever its origin, the entity thus produced, however unlifelike, gave every impression of being under independent control or even of being capable of independent action — independent, at least, of the medium's conscious mind. Richet observed the same thing with regard to the materialization phenomena of Eusapia Palladino:

> The ectoplasmic arms and hands that emerge from the body of Eusapia do only what they wish, and though Eusapia knows what they do, they are not directed by Eusapia's will; or rather, there is, for the moment, no Eusapia. [130]

We may interpret this how we will. It is noteworthy that Richet himself did not think that it necessarily had to be interpreted as an indication of a separate agency:

> The personalities that appear in the course of experiment do not seem more conscious of themselves than those which manifest by automatic writing. They seem to pertain more or less to the conscious or unconscious fancy of the medium.

Setting aside, as I think we must, any suggestion that every recorded instance of materialization was, in fact, an imposture, we are forced to accept a degree of reality in the materialized entity. We have to admit the possibility of the production of a certain substance that can be moulded into the likeness, however rough-and-ready, of a human being, and that occasionally bears a close resemblance to an individual known to the observers.

The best recorded cases are those in which scientific curiosity, rather than emotional desire, is the motivation; and this has very important implications:

★ it could be read as an indication that the spirits are willing to provide proof of their existence, and to this end are happy to co-operate with hard-headed psychical researchers just as with credulous believers; or

★ the subconscious mind is not concerned with motive or purpose, but is willing to do whatever is asked of it, which may be the same as saying that it is infinitely suggestible and at the same time utterly undiscriminating.

It is evident that this type of entity — if indeed it qualifies for that label at all — is in many respects very different from those others we have been considering. On the face of it, nothing could be further from a hallucination than this ooze from a medium's navel, which gradually resolves itself into a far from convincing shape to be peered at by a group of sceptical scientists. Even if we look at the materialization phenomenon through a Spiritualist's eyes, it seems a very different thing from the apparition of a Virgin or the conjuring up of a spirit from the abyss.

And yet the distinction is not an absolute one. For these materializations have one very important feature in common with some, at least, of our other entities: they are subject to the control of, and are perhaps created by, the subconscious mind of the central figure in the incident, be he medium or visionary or passive percipient. And this is true whether we take the Spiritualist view, and suppose that that mind has temporarily been taken over by agents from the spirit world, or if we think that it is responding, by some subtle psychological process, to a more mundane stimulus. Dr Charles Ségard, one of the investigators who accompanied Richet in the sittings with Eusapia Palladino, came to this conclusion:

It appears to result:

1   That the majority of these phenomena are undeniable, and that they exclude all suspicion of fraud.

2   That it is our inherited traditional beliefs, sometimes dulled but rarely altogether extinguished, which, aroused once more by these observations, have probably led to the current interpretationof the unaccustomed facts and inclined us to believe in the intervention of 'spirits' different from ourselves, although still too much resembling us.

3   That it is we ourselves, whether as subjects or as spectators, who lend to these 'apparitions' an identity of their own; on the whole it would be more natural, more logical, to see in them only a duplication of our own physical and moral personality. [140]

2.8 HALLUCINOGENIC ENTITIES

Probably the most notorious example of a hallucinated experience is that of witches, who believed that they went on flights, mounted on broomsticks or otherwise, and participated in ceremonial rites. The literature of medieval witchcraft is rich in such accounts, and though many of them were given under pressure of torture and other kinds of stress, there is no reason to question that a good many such confessions were made in total sincerity — that is, the accused person really did believe the experience had occurred.

The process is clearly described in this account which dates from seventeenth-century Germany:

> A certain priest of our order entered a village where he came upon a woman so out of her senses that she believed herself to be transported through the air during the night with Diana and other women. When he attempted to remove this heresy from her by means of wholesome discourse, she steadfastly maintained her belief. The priest then asked her, 'Allow me to be present when you depart on the next occasion.' She answered, 'I agree to it and you will observe my departure in the presence (if you wish) of suitable witnesses.' Therefore, when the day for the departure arrived, which the old woman had previously determined, the priest showed up with trustworthy townsmen to convince this fanatic of her madness. The woman, having placed a large bowl, which was used for kneading dough, on top of a stool, stepped into the bowl and sat herself down. Then, rubbing ointment on herself to the accompaniment of magic incantations, she lay her head back and immediately fell asleep. With the labour of the devil she dreamed of Mistress Venus and other superstitions so vividly that, crying out with a shout and striking her hands about, she jarred the bowl in which she was sitting and, falling down from the stool, seriously injured herself about the head. As she lay there awakened, the priest cried out to her that she had not moved. 'For heaven's sake, where are you? You were not with Diana and, as will be attested by these present, you never left this bowl.' Thus, by this act and by thoughtful exhortations, he drew out this belief from her abominable soul. [149]

By way of illustrating the stereotyped character of the phenomenon, here is an example from another country and another period, sixteenth-century Italy:

> Dominus Augustinus de Turre, of Bergamo, the most cultivated physician

of his time, told me that when he was a youth at his studies in Padova, he returned home one night about midnight. He went to look for the maid and finally found her lying in her room, supine upon the floor, stripped as if a corpse, and completely unconscious, so that he was in no way able to rouse her. When it was morning, and she had returned to her senses, he asked her what happened that night. She finally confessed that she had been carried on the journey.[65]

The stereotyping covered not only the experience itself but also the preparations: in this respect the practice resembled magical conjuration — the performance of a specific ritual would result in a specific effect. Evidently anointing the body with a hallucinogenic substance was part of the ritual, as also were the magical incantations.

The parallel with shamanistic practices in the primitive cultures of our own day is inescapable. In a wide variety of cultures in various parts of the world it is regular practice for people to have recourse to hallucinogens in order to induce effects that are similar in many respects to those reported of the medieval witches.

Procedures vary. Typically, the shaman acts as doctor/magician/counsellor/detective to his community, and he may be responsible for guiding the participants in the experience. Sometimes it is he alone who has the experience, acting on behalf of those who have consulted him or sought his help. The experiences vary considerably, but it seems probable that the variations are cultural rather than fundamental — which could be rephrased by saying that we all get the variety of religious experience we ask for.

Whoever subjects himself to the experience, and with whatever motive, what results is the consequence of the bio-chemical properties of the substance eaten, drunk or smoked on the one hand and, on the other, of the pre-existing cultural expectations of the subject. Michael Harner, studying the banisteriopsis rituals of the Zaparo communities of eastern Ecuador, listed the effects as follows:[65]

1   The soul is felt to separate from the physical body and to make a trip, often with the sensation of flight.

2   The subject sees visions of jaguars and snakes and (to a much lesser extent) other predatory animals. (These particular animals clearly have a cultural significance for the Zaparo, due to their prevalence in their environment.)

3   He has a sense of contact with the supernatural, whether with demons or — in the case of those exposed to Christian missions — with God, heaven and hell.

4   He has visions of distant persons, cities and landscapes. The Indians interpret these as visions of reality; that is, as clairvoyant experiences.

5   He has the sensation of seeing the detailed enactment of recent unsolved crimes, particularly homicide and theft. This is evidently the result of expectation: what it shows is that the experience is not altogether uncontrolled, but is *directed* towards the purpose the subject had in mind.

These results show that what the drug is doing is allowing the subject's mind to enter a freer, less inhibited state, in which his wishes and expectations find expression in a context determined by his cultural background. He, of course, believes that it is the drug that is doing the work, whereas the drug is merely releasing the brake imposed on his mind by the need for everyday social constraint.

The fact that the practice is used for different purposes in different cultures will cause the experience itself to differ accordingly; so will the archetypal elements such as magical jaguars, totem snakes, wise gods and ancestors, and other figures drawn from the prevailing belief-system.

Even if this were all, it would show that the hallucinogenic experience is relevant to our inquiry; for we can hardly help noting similarities common to this kind of experience and some of our entity-sightings, showing that hallucinogens provide at least an alternative route to the visionary experience. But there are findings that come even closer to the subject of our study. Claudio Naranjo conducted an experiment in Santiago, Chile, in which thirty-five volunteers took the hallucinogenic drug *yagé*, or *ayahuasca*, without knowing in advance what effects had been reported of it by anthropologists. [109]

He found effects that are quite different from those brought about by mescalin or LSD; though they induced a trance-like state, it was characterized by a substantial mental alertness. Most commonly, the subject would visualize with closed eyes images he did not mistake for reality, but which were of very great intensity and accompanied by strong emotional responses. The sensation of flying was just one of the physical illusions; another was of seeing the world from a height, generally looking down from a flying position.

Many animals were encountered, and many religious archetypes and symbols. But most relevant to our study,

> — five persons saw the Devil or devils
> — three saw 'angels'
> — three saw the Virgin Mary
> — two saw Jesus Christ

Chile is, of course, a strongly Catholic country, so if visions of any kind are seen, they would be likely to be those which were reported. The conclusion is self-evident: this kind of hallucinogen creates a state of mind that is favourable, not simply to hallucinatory experiences, but to experiences of a specific kind, in which images of personal and cultural significance to the subject will manifest themselves. In other words, hallucinogens could be one of the ways in which percipients come to have the kind of entity-seeing experience this inquiry is concerned with. The only question at issue is whether they *cause* the experience, or simply *enable* it to occur — a question which has already risen in respect of other kinds of experience, and which we shall discuss in the concluding part of our study.

2.9 SPONTANEOUS AND EXPERIMENTAL CASES: A SECOND ANALYSIS

By the end of Part One, we had established that a great number of people had seen a wide variety of entities under a no less wide variety of circumstances. Despite their many differences they also had several characteristics in common, though the only thing they *all* had in common was that none of them, however real it appeared to be, seemed to be real in the sense that you who read this and I who write it are real. The question inevitably rose, whether all these reported entities are specific types of phenomenon in their own right, or different versions of a single phenomenon varying according to circumstances.

At that time it was clear that we did not have sufficient evidence to answer that question, let alone to proceed to an explanation for the phenomena. In Part Two, therefore, we have examined various kinds of related experience to see if they throw any helpful light on the spontaneous phenomena. We can now take stock of the evidence we have accumulated:

★ From our study of *dreams*, we learn that there exists within the human mind an infinite storehouse of recollected material, and also an autonomous instrument, which we labelled the 'producer', capable of manipulating and deploying that material in highly elaborate scenarios. This creativity seems to serve no biological purpose, but may conceivably serve a psychological one as some kind of mental housekeeping or therapy; yet though a profound purpose seems sometimes to be indicated, most often the presentation seems pointless and even frivolous. All of which makes it harder to understand why the 'producer' exists at all, and why he should have been endowed not only with such extraordinary powers of creativity but also with access to information which is not available to the conscious mind.

★ From *hypnagogic/hypnopompic experiences* we learn that in certain mental states (in this case, the threshold between sleeping and waking) the mind has presented to it highly detailed images which seem to have no counterpart in reality. The fact that a specific physiological state is associated with a specific category of image experience seems to imply a straight cause-and-effect relationship, but why this association should exist is as obscure as is the purpose of the process.

★ From *hallucinations* we learn that the subconscious mind can present illusory visual experiences to the waking mind that are so lifelike as to impose total conviction even though the percipient may be perfectly well aware that they are illusory. Moreover, these images are not simply imposed on reality but adapted to the context in which they are 'seen', demonstrating once again the resourcefulness of the 'producer'. While this type of experience is often associated with sickness and other special states, this is by no means necessarily the case, and all we can conclude is that fevers and so forth somehow enable the process to occur more readily.

★ From *doubles and doppelgängers* we learn that a percipient may have an illusion of seeing a perfect replica of himself, which acts quite autonomously, which seems to have access to information denied to his waking self, and which generally manifests with some purpose. There are indications of an association with migraine and other physiological states, but once again this is by no means necessarily so, and so we must suppose it to be an enabling rather than a causal function.

★ From *companions and counsellors* we learn that some people are able to create, or to invite into their company, entities who proceed to play an intimate part in their lives over a period that may cover many years. In general, these entities seem to be strangers to the percipient, but there are cases in which there are indications of an entity similar to that seen in haunting cases; it is possible that these should be treated as a separate category. If the companions are created by the percipient, this would seem to be an unconscious process; but because the percipients are generally young children this is not a very meaningful distinction, for it is hardly possible to say whether a child's desire for company should be thought of as conscious or unconscious. A similar kind of experience is had by adult percipients, but this is not usually on a long-lasting basis and again perhaps it is wrong to classify the two types of phenomena together: both, however, seem to occur in response to a wish or need on the percipient's part.

★ From *apparitions* we learn the virtual impossibility of attributing all entity experiences to processes within the percipient's mind. In apparition cases, visual replicas of persons, who may or may not be known to the percipient, manifest in his presence, and display more often than not an awareness of both the percipient and the physical environment. They frequently indicate an ostensible purpose, which is more often related to the apparent's circumstances than to those of the percipient. Information is often provided which could by no normal means be known to the percipient, providing persuasive evidence that an external agency is at least partially responsible. On the whole, these characteristics seem to require an explanation that presupposes co-operation between the percipient and the agent, on a subconscious level as regards the first and probably in most cases the second also.

★ From *hauntings* we learn that apparitions of people known to be no longer alive are seen recurrently, in or close to a specific location, often over a period of many years and by several independent percipients who are more often than not total strangers to the apparent. Sometimes there are indications of purpose. The evidence for an external agent, independent of the percipients, is so strong as to be overwhelming; though again we may suppose some kind of interaction to occur, if only to explain why the apparition is seen only intermittently.

★ From *visions* we learn that some percipients have encounters with entities who are not known to them personally, but whom they know about in a general way: they are most often eminent figures in the percipient's religious belief-system. These apparitions are in many respects different from those in the two previous categories. They appear to the percipient(s) alone, even though other people may be present; they generally pay not one but a succession of visits in the space of a few months, but do not return thereafter; if several people together see the same entity they describe it in much the same terms, but if independent percipients claim to see an entity purporting to be the same individual (for example, the Virgin Mary), their descriptions are apt to be widely divergent. This type of experience is particularly associated with children between the ages of nine and fifteen, more often female than male, almost always of Catholic upbringing, and generally of simple or deprived social status. Unlike any other kind of entity-sighting, this form of experience has been seen as having great significance, not simply for the percipient but for a much wider community, and attracts enormous public interest as a result. The fact that this in turn seems to generate further paranormal happenings serves to increase the significance accorded to visionary experiences. There is evidence to show that visions are seen in greater numbers when there are more than average reports of unidentified flying objects: this simple correlation may be our first indication of a much larger matrix of relationships that could prove fundamental to the entire range of phenomena.

★ From *demonic entities* we learn that some percipients encounter entities whom they identify as beings that figure in their belief-system, though there may be no other evidence for the existence of such beings. The possibility remains that they may be all-purpose spirits of no specific character, assuming a specific form for the purpose of visiting the percipient, which of course will be one that has meaning to him. Some very elaborate scenarios have been devised on this basis, but they rest on assumptions for which no reliable evidence exists.

★ From *UFO-related entities and Men In Black* we learn how, in a comparatively short space of time, a new type of entity can come into being in close association with a wider cultural manifestation. If it should eventually turn out that we are really being visited by extraterrestrial aliens, and that the entities seen in association with UFOs *are* those aliens, then of course they will no longer form part of the subject-matter of this inquiry. However, there are good grounds for doubting whether this is the case, and even if it should be so, it seems likely that a good many of the alleged extraterrestrial entities *are* apparitions of some kind, in a form compatible with current preoccupations. This is borne out by the way the great majority of the entities encountered in contact cases reflect, in their 'messages', current human fears and expectations. Significantly, too, a very high proportion are seen only by an individual percipient, who is indicated as having been chosen as a privileged contact or envoy.

★ From *folklore entities* we learn that stereotyped entities, which do not seem to have any origin in fact, can serve as a pattern for real-life entity-experiences for which substantial evidence, including multiple testimony, is forthcoming. The ostensible implication is that a prevalent stereotype can find physical expression in a totally lifelike manner.

★ From the *imaginary abduction experiments* we learn that, under hypnosis, people can describe fictitious 'experiences' so elaborately and vividly as to be indistinguishable from descriptions given by witnesses who claim to have had *real* experiences; and this despite the fact that they seem not to have had access to the requisite sources of information to enable them to construct such complex scenarios. ESP may provide an explanation, but a very unsatisfactory one; another possible hypothesis would suppose that both sets of percipients, the 'real' and the imagining, are offering scenarios embodying similar archetypal material, though the stimulus provoking the experience differs from one set to the other and may cause differences in the way the account is given, though not in the account itself.

★ From *age regression hypnosis* we learn that the mind, in the hypnotic state (and this may apply to other alternative states of consciousness also), will, if confronted by an anomalous element when presenting a scenario, find ways of absorbing that element into the scenario instantly and spontaneously: age-regression subjects account for the presence of the anachronistic hypnotist in much the same way as the dream 'producer' accounts for the sound of an alarm clock or the feeling of cold. This experimental evidence confirms and extends our awareness not only of the 'producer' 's existence but also of his infinite creative capability.

★ From the (as yet, unique) *'Ruth' case*, we learn that an entity which appears to act autonomously, and to appear to the percipient independently of his wishes, is nevertheless a figment of his imagination, and capable of being controlled or dismissed by his conscious mind. We also learn that a percipient can learn to create such hallucinations on demand, and that even so they will, having been created, conduct themselves autonomously and in a totally lifelike manner. Yet again, we seem to be confronted by an example of the 'producer' 's remarkable skill.

★ From *experiments in projection* we learn that it is possible for a person to project an image of himself to a distant place, where that image can be seen by one or more persons who may or may not be known to the agent. The experience in many ways suggests the transmitting end of an apparition experience; projected images share with apparitions the ability *not* to be seen by someone present alongside the percipient. The agent may be awake or asleep; if awake, he may or he may not have the experience of taking part in the 'visit'; but if he does, he may acquire knowledge about a place that he has never seen in his waking life. At the same time, his image will conduct itself in that place as though aware of its surroundings, and react to the percipient and even on occasion communicate with him.

★ From *magical conjurations* we learn that certain people, if they work themselves up into a suitable emotional state by means of ritual practices and other stimuli, can come to believe they have conjured up entities whom they generally suppose to be beings from a parallel, non-human plane of being.

★ From *thoughtforms* we learn that some people may, by strenuous acts of will, be able to create non-physical beings which are, nonetheless, sufficiently material to be seen on occasion by others. Once created, these entities act autonomously and may develop a character of their own.

★ From *spirits of the dead* we learn that dead persons may be able to project an image of themselves back to Earth, sometimes making use of material derived from a medium's body. This interpretation of the phenomenon (which does seem on occasion to be real enough) is not, however, the only one available, and there are many cases for which it is very unsatisfactory. Once again, however, the evidence points towards a physical consequence from a mental process, whether or not we presume the activity of an external agent.

★ From *hallucinogenic entities* we learn that certain drug-induced states are favourable to the illusion of seeing entities who share many characteristics with those of religious visions and folklore. This evidence, again, provides an example of a physiological process combining with cultural imagery to create an experience which is of overwhelming personal significance for the percipient.

It is still far from clear whether we are dealing with a single phenomenon, or different forms of a single phenomenon, or several phenomena that share certain characteristics. Clearly, there *are* shared characteristics: equally clearly, there are differences. However, I suggest that nothing we have learned rules out the possibility of some kind of field hypothesis, and even if it should not be possible to devise a hypothesis that has more than a limited applicability, even that will be of some value.

In the following and concluding part of this inquiry, we shall consider some further findings, of a more speculative kind, which should help us answer three questions which, now that we know *what* occurs, have become the most urgent:

1   *How* does the entity sighting experience occur?

2   *Why* does the entity sighting experience occur?

3   Why does the entity sighting experience take the *form* it does?

PART THREE

EXPLAINING THE ENTITY
EXPERIENCE

INTRODUCTION

Now that we have surveyed a reasonable cross-section of the evidence for entity-sightings, we are in a position to review suggestions as to how they come about, and to consider what conclusions can legitimately be derived from those suggestions. What we are looking for, in this third part of our inquiry, are the mechanisms and processes that might cause people to see, or think they see, entities.

No proposals will be viable unless they can account for the many paradoxical features reported by entity percipients. Among these are:

★ Entities are not real in the accepted sense of the term. This is shown, for example, by their defiance of physical laws; by the fact that the apparent may be known to be in another place, or dead; by the fact that some of those present see the entity while others do not. At the same time, many entities appear to have reality of a sort, for they are seen by more than one person, and on more than one occasion.

We have to ask whether this can be explained by reference to suggestion, collective hallucination and so forth; or whether we should give serious consideration to the possibility that entities do exist on some alternative plane of reality.

★ Entities are capable of being so lifelike that even when the percipient knows at the time of his sighting that what he sees has no reality in the conventional sense, he will continue to see it in defiance of what his reason tells him.

We have to ask if this is because he is seeing something that, while not real in the accepted sense, does have some alternative mode of existence; or have his senses somehow been by-passed and deceived by some agency other than his conscious mind?

★ Entities frequently convey information that is not available to the percipient.

We have to ask whether the percipient has acquired the information paranormally, and attributes it to an imagined entity (which he himself creates) as a way of hiding from himself the method by which the information was acquired; or does the entity originate with an agent who does possess that information?

★ Many entities show a considerable capability for autonomous activity; they display awareness of their surroundings, and respond to the percipient even to the extent of intelligent communication.

We have to ask if this appearance of autonomy is genuine, or simply part of a display mounted by the percipient's own imagination. In answering, we must take into account those cases in which the entity has been deliberately projected by an agent, and as a result of which the agent obtains an 'entity's-eye' view of the experience, including knowledge of the previously unknown location in which it occurs.

★ Entities often reflect cultural stereotypes, such as extraterrestrial aliens or women in white.

We have to ask ourselves whether this is because they are imaginary constructs created by the percipient's own imagination using the currently available stereotypes, or whether both the stereotypes and the entities relate to some kind of common reality.

★ Entities are often controllable by the percipient, who can direct or dismiss them, and in some cases cause them to appear in the first place.

We have to ask whether the obvious explanation — that they are mere figments of his imagination — is really adequate: how does it account for cases in which an external agent seems to be responsible? Could it be that some entities, at least, are externally originated, but able to manifest only with the co-operation of the percipient's own subconscious mind; and when that support is withdrawn, the manifestation can no longer be sustained?

★ Entities frequently appear to people who are in a special state — relaxed and passive, as in the hypnagogic state; in sickness, as in fevers which induce delirious hallucinations; in drug-induced states which lead to visionary experiences; in mystical states which also lead to visions; or in a crisis situation, when entities may bring help or comfort.

We have to ask whether it is *necessary* to be in some such state to experience the sighting; and if so, whether the physical or psychological circumstances *cause* the event to occur, or simply *make it possible* for it to occur.

★ Entities frequently display a purpose, but this may relate either to the needs of the apparent — for example, when he is involved in an accident — or to those of the percipient — as when it is he who is in need of help.

We have to ask whether explanations in terms of ESP in the first case and 'wishful thinking' in the second case are really satisfactory, or if we should propose a model that accepts some kind of externally originating agency?

There are, of course, many other puzzling aspects of the entity phenomenon; but if we can devise a model that satisfactorily resolves the foregoing, we should expect to find that it resolves many of the others also.

3.1 THE COLLECTIVE UNCONSCIOUS

Our survey of the entities that people claim to see shows that they fall into three main categories:

★ individualized entities, such as the apparition of Aunt Agnes

★ stereotyped entities, such as a vision of the Virgin Mary

★ generalized entities, such as an extraterrestrial alien

These appear to be quite different and distinct categories; but the differences may not be so great as they appear to be — they may form no more than specific variations on what is fundamentally one and the same phenomenon. If this is so, then we should not allow ourselves to be distracted by the superficial characteristics, but look for some kind of undifferentiated proto-entity, like the characterless Barbie-doll my daughter used to dress up in whatever clothes she chose. Just as she, in so doing, conferred a specific identity on a mass-produced zombie, so it may be that our entity percipients bestow, on the proto-entity, the features of Aunt Agnes or the imagined characteristics of the Virgin or the extraterrestrial.

We have already encountered something like this hypothesis, when we were considering folklore entities in Chapter 1.12. The theosophist Gardner presented an account of fairies that saw them as essentially amorphous creatures, which take on the likeness of fairies, gnomes and so on in their dealings with humankind. Whatever we may think of this particular suggestion, the process it incorporates is one that is plausible and very attractive as a way of accounting for the puzzling confusion of similarities and differences entity cases present.

But does there have to be a proto-entity, in any material sense, to provide the basis for the sighting? The evidence for such a being is, after all, purely circumstantial. For alternative models we must, of course, look to the psychologists: and particularly to C. G. Jung, whose brilliant concept of the *collective unconscious* offers us an approach to the entity problem that effectively accounts for many of its paradoxical features.

Crudely summarized, Jung's suggestion is that, beneath the individual awareness of the universe that each of us acquires in the course of our lives, we share with our fellow-creatures a more fundamental set of patterns and paradigms; these are part of our genetic inheritance, rather than acquisitions made in the process of individual education or cultural conditioning. The individual items of shared 'information' he named *archetypes,* which form a kind of symbolistic currency that can be passed from one person to another enjoying universal acceptability.

For Jung, a working psychoanalyst, archetype-spotting was a helpful diagnostic aid. If a patient told him she had been dreaming about snakes or eggs, this would be a clue as to what concealed preoccupations were working within her. But the question of what value such concepts have outside the analyst's consulting room remains unresolved. The temptation to apply the hypothesis more widely is strong, though; and we must certainly ask if it has any viability in the context of our inquiry.

Jung himself was one of the first to see if his concept could work on a mass phenomenon as opposed to individual cases. Almost at the very end of his life, the 'flying saucer' phenomenon burst upon the world, and Jung with remarkable perception saw its significance for the behavioural scientist. He also saw that it provided a unique opportunity to try out his theory of archetypes. He suggested that the saucer shape reported by so many witnesses corresponded to one of the fundamental archetypes, the mandala, which he had already discussed in many previous writings. [78] Now, in his study of the flying saucer phenomenon, [79] he argued that UFOs are 'a modern myth of things seen in the skies'.

It is important to note that though it was the psychological aspects of the phenomenon that interested him, Jung did not insist that flying saucers are purely psychological in their nature. His interpretation of their significance, as he points out, remains valid whether or not the saucers have any basis in reality:

> Thus there arose a situation in which, with the best will in the world, one often did not know and could not discover whether a primary perception was followed by a phantasm, or whether, conversely, a primary fantasy originating in the unconscious invaded the conscious mind with illusions and visions . . . In the first case an objectively real, physical process forms the basis for an accompanying myth; in the second case an archetype creates the corresponding vision.

That is to say, it is not clear whether the percipient sees something real, and invests it with the trappings of the archetype, or whether his desire to see the archetype stimulates an imaginary experience. So far as the psychologist is concerned, this does not matter too much, for it is not so much *what* he sees, as *why* he thinks he sees what he thinks he sees, that is the province of the analyst. So, though the rest of us may continue to feel a certain curiosity as to what is *really* going on, let us set that curiosity aside for the moment and continue to explore the psychological implications.

A fundamental question many of us will feel like asking is, why do people want to see archetypes at all? Since Jung devoted many thousands of pages to this question, it is impossible to give a very satisfactory brief answer; but here is a passage in which Jung gives some indication of the processes at work:

> In the individual, such phenomena as abnormal convictions, visions, illusions etc., only occur when he is suffering from a psychic dissociation, that is, when there is a split between the conscious attitude and the unconscious contents opposed to it. Precisely because the conscious mind does not know about them, and is therefore confronted with a situation from which there seems to be no way out, these strange contents cannot be integrated directly, but seek to express themselves indirectly, thus giving rise to unexpected and seemingly inexplicable opinions, beliefs, illusions, visions and so forth.

If Jung is right, then our entity percipients are each of them working out an individual problem in his own way, and expressing it in terms of an archetypal being appropriate to this problem. But how does this work on a global scale?

> Projections have what we might call different ranges, according to whether they stem from merely personal conditions or from deeper collective ones. Personal repressions manifest themselves in our immediate environment. Collective contents, such as religious, philosophical, political and social conflicts, select projection-carriers of a corresponding kind.

This leads to the question: when an individual percipient sees an entity, does he do so as an individual — seeking a solution for his individual problem — or as a member of the community — seeing a symbol of universal significance that expresses the communal *angst* of his place and time?

This is where Jung's hypothesis works superbly: for it proposes that the percipient undergoes his experience *both* as an individual *and* as a child of his time. And the result is that he sees an entity that has a communal significance — the Virgin Mary, an alien visitor — but with specific attributes that relate to his own situation — the Virgin gives a message of personal comfort, the alien shares his preoccupation with ecology or The Bomb.

If any Jungian scholar should accuse me of debasing the master's profound insight by applying it so simplistically to the question of what makes a group of schoolchildren think they have been visited by the Queen of Heaven, I would have to plead, if not guilt, then certainly ignorance. For despite my admiration for Jung's theory, I am not really clear exactly how he conceived it as operating in practice. His archetypes — those which are in the form of human figures — are notable for their generalized form — 'the mother', 'the maiden', 'the wise old man', 'the earth mother', 'the trickster', and so forth. It would not be difficult, even among the entity cases quoted in this survey, to find instances in which the entities can be matched against such generalized concepts. For example, the 'Men In Black' are easily identified as 'tricksters', while the visionary Virgin seems to combine some of the attributes of the 'maiden' (it is significant that most visionaries see her as far too young to be the mother of Jesus) with others that pertain to the 'earth mother' archetype. But these are interpretations done in retrospect: the entity having been seen and described by the percipient, we can set to work to match it against our check-list of archetypes. But the question still remains: was the entity created with the archetype as a blueprint? Are we to assume that each and every one of our entity visions is the consequence of an individual working out his personal situation in terms of the collective unconscious?

I think we can be fairly confident that this is *not* the case. All we are justified in saying is that Jung's archetypes are one of the many sources from which the 'producer' obtains his inspiration. Indeed, we may be wrong in thinking that these archetypes exist in the kind of absolute sense Jung seems to have had in mind; perhaps we should conceive them as *directing tendencies.* That is to say, when a Bernadette sees a helpful entity, we should think of that entity not as an approximation to an archetype that pre-exists like some kind of Platonic Ideal, but rather as a creation whose form has been determined — in part at least — by the fact that *Bernadette's needs are not unique to her alone, but shared with the rest of us to a greater or lesser extent. Consequently, they find expression in attributes whose validity we all either consciously recognize or unconsciously feel*: to this extent they are archetypal.

Perhaps I may seem to be splitting hairs; but if so, it is because I am anxious, in this dauntingly complex business, to present as much as possible of the material in the form of known or easily understood processes, rather than as conjectural concepts forming part of a specific theory-system. We can benefit from Jung's thinking without necessarily committing ourselves to the theoretical structure he proposed.

I do not think there is any doubt that Jung's thinking is of very great relevance to our inquiry. We do not have to accept the universal validity of the collective unconscious and archetype hypotheses to appreciate the value of these concepts in illuminating the processes at work in the sighting of an entity. By directing our attention to the puzzling combination of 'public' and 'private' elements displayed in so many entity-sightings, and by showing how that combination can be plausibly accounted for, Jung's ideas are a major step towards answering one of our three crucial questions: Why does the entity sighting take the form it does?

3.2 THE IMAGE-BANK HYPOTHESIS

The notion of a storehouse containing all knowledge, of the future as of the past, is one that occurs in many different contexts. Occultists believe in 'the 'Akashic Records', while Christians traditionally expect St Peter, at the Gate of Heaven, to have a book in which their life-stories are recorded in dismayingly intimate detail. For philosophers grappling with the notion of time, and for psychologists concerned with the anomalies of memory, the concept is virtually indispensable. For the purpose of our inquiry, too, it is a convenient device that is particularly applicable to the perplexing problem of apparitions.

The question of memory is a very complex one. We know that we have access to an almost infinite range of material — the most trivial incidents of our distant past, apparently without any emotional or other significance for us, suddenly emerge in the course of a dream, or evoked by a smell, a taste or a reference in conversation. We know that under hypnosis, the 'amnesic barrier' (Hilgard's phrase) can be breached and we can be shown to have retained visual images of scenes we certainly made no conscious effort to record — the registration numbers of cars is the best-known example. There seems to be no limit to the amount of information that is potentially available in this way.

Is all this information stored in the human brain? Neither physiologists nor psychologists can give us a definite answer. It is true that certain forms of physical probing seem to evoke submerged memories, but it does not follow that the probe has reached the storage point; it may have done no more than stimulate the information channel to the store, as switching on a radio enables the listener to obtain information that derives not from the radio set but from a distant broadcasting station.

It is difficult to avoid analogies drawn from our own information technology, dangerous as such models are. A distinction has been drawn, for example, between short-term and long-term memory: is it right to compare this with the practice of an office that keeps recent material in file cards on the premises, while older records are stored on microfilm in the vaults?

So far as individual memory is concerned, it is logical to suppose that each of us has his own memory-bank containing all the memories we have personally experienced. We may add to this, material we have inherited, which could even include the archetypes of the collective unconscious, discussed in the preceding chapter, that Jung seems to have thought are genetically transmitted. But how are we to account for the fact that, from time to time, one of us will display awareness of information that lies outside our personal experience, and to which we do not seem to have had access by any of the normal channels of communication? To take a specific instance, how are we to account for the fact that George, seeing an apparition of his Aunt Agnes that is subsequently established as having occurred at the same moment as she was involved in an accident, sees her wearing the clothes she was indeed wearing at the time, clothes he did not know she possessed? And how do we account for the fact that Aunt Agnes, if she is responsible for projecting her image to George, knows where to find him?

This kind of occurrence is generally attributed to extra-sensory perception. It is true that many people, because the evidence for ESP is only circumstantial, prefer to attribute the supposed instances of its occurrence to chance or coincidence: but the majority would probably agree that some such capability is the most logical explanation for innumerable incidents. When a wife has the feeling that her husband is in danger, for example, and it subsequently turns out that he had a boating accident at the time, telepathy — a message passed from his subconscious mind to hers — does seem both the simplest and the most logical explanation.

The phenomenon of psychometry makes greater demands on the ESP explanation: when someone holds an object in his hands and proceeds to describe its previous owners, or incidents associated with it, we cannot easily resist the conclusion that the information is somehow 'contained' in the object, or that it carries with it an 'aura' that records everything that has ever happened to it. But if we propose such a hypothesis, we are well on the way to proposing an image-bank in which *all* such information is contained.

Information acquired from the dead presents us with another puzzle: we have to suppose that the dead have survived, and that we can establish communication between our minds and theirs. This may be the case; but there are reasons for doubting it, of which one of the crudest is also the most convincing — the fact that rarely, if ever, does communication with the supposed dead bring to light information that could not have been known to the dead person when living. So again, the image-bank hypothesis offers an alternative.

Here are two cases in which the image-bank hypothesis offers a plausible *modus operandi*. The first is very similar to that of the lady climber whose rescue by an oriental figure we considered earlier in connection with apparitions; perhaps there, too, the image-bank hypothesis offers an explanation.

During World War One a Canadian prisoner of war in Germany managed to escape, and eventually reached a point at which he arrived, at night and during a snowstorm, at a crossroads where one road led to Holland and probable safety, the other to a near-certain capture. He hesitated, then opted for one of the roads. Suddenly there appeared before him the figure of his brother, utterly clear and lifelike, who said to him, 'No, Dick, not that way. Take the other road, you damned fool!' As a result the escaping man took the other road and reached safety. [127]

At the time, the brother, an officer, was probably asleep in England. He had no idea of his brother's predicament, and could not possibly have advised him which road to take. In short, the brother seems to count for nothing in the whole affair, and one can easily assume — as does Louisa Rhine, who reports the case — that the entity-brother was a dramatization created by the percipient's subconscious mind, a convenient way of expressing the warning, just as the 'bodhisattva' had been for the lady climber.

But this is not in fact necessarily the correct interpretation. Suppose the subconscious mind of the brother, then sleeping back in England, became aware of the percipient's dangerous situation; his concern for his brother could have caused him to intervene, in the form of what amounts to a crisis-apparition.

In this case, it is reasonable to suppose that *nobody* possessed the information — that is, there was no living person who knew both which was the safe road, and that the escaper was in danger of choosing the wrong one. Whoever/whatever guided the percipient, it was someone/something with access to more than commonly-accessible knowledge. It could have been the percipient himself, but we then have to ask why he went to all the trouble of dramatizing his brother; after all, there are dozens of cases on record in which people have been warned by simple premonitions, or even by imaginary disembodied voices. The possibility that it really was the brother who was the agent, though admittedly it seems to complicate the matter unnecessarily by introducing an extra person, is nevertheless more plausible. *But even if we have to leave open the question of whether the agency was internal or external, this much is certain: either way, access to some external source of information occurred, and something like the image-bank hypothesis is needed to account for it.*

Here is another case, which differs from the foregoing in that the information was available by normal means (if you are prepared to accept telepathy as normal!) and yet the image-bank hypothesis may still be more plausible:

> An American dentist was preparing to extract a tooth from a patient unknown to him, and was about to give him a Novocain injection, when the image of his mother, who had died twelve years previously, appeared in the doorway and said to him, 'No, John, no. He thinks the narcotic will kill him. He wants to die in your chair so you will be blamed and his family will not lose his life insurance'. The dentist asked the patient if he was unable to take Novocain; the patient replied that though he didn't know this to be certain, it was a fact that he had suffered ill effects from a previous injection. [127]

Here, of course, the patient did have the information and knew his own intention, so it is tempting to take the obvious course and suppose that the dentist acquired the information from his patient's mind by some form of ESP — if one accepts that ESP is a viable way of acquiring information. But once again, why involve the dentist's dead mother? It seems an absurdly complicated way for the dentist's subconscious mind to warn him not to give the injection, except on the hypothesis that it really *was* the mother who gave the warning, seeing her son on the point of making a terrible mistake. If so, how did she get the information? Once again, the image-bank hypothesis offers a plausible explanation.

Yet another area in which knowledge seems to be acquired paranormally is in connection with the curious phenomenon of 'out of the body' experiences, also known to some as 'astral travelling'. One of the characteristics of this kind of experience is the obtaining of knowledge of distant places by the OBE subject whose physical body has certainly not left his own home: to call this 'travelling clairvoyance' is to do no more than give it a label. One OBE subject said of his experience: 'I realized that I was seeing, not only things at home, but in London and in Scotland, in fact wherever my attention was directed'. [163]

The image-bank hypothesis would help with some of the problems we have encountered in this inquiry; for instance, how Lawson's volunteer imaginary abduction subjects (Chapter 2.1) came up with narratives so similar to those recounted by the supposedly real abductees. It would help us account for the bizarre circumstances we shall be noting later on when we consider the perplexing correspondences between science fiction and certain categories of anomalous experience. And it provides an alternative way of looking at the similarities of experience that Jung believed demonstrated the existence of a collective unconscious. (The two concepts are not, of course, irreconcilable; there is no reason why Jung's archetypes should not be stored in the image-bank!) Similarly, the image-bank hypothesis is compatible with the morphic resonance of Rupert Sheldrake's theory, which suggests that organisms are somehow 'aware' of what happens to other organisms of their type. His morphogenetic field is in this respect another metaphor for our image bank.

What form the image-bank would take is immaterial, as are the conditions under which people obtain access to it. But it seems at least a reasonable hypothesis to suggest that the conscious mind, in its normal waking state, does not have access to this store, but that the subconscious mind does: and further, that the presence of certain conditions — a state of altered consciousness such as dreaming or hypnosis, a personal crisis, a delirium induced by fever, a drug-induced state of enhanced perception, and so on — could provide a means of by-passing the normal restrictions and gaining access to this resource.

I am aware that this is sheer speculation, albeit of a kind which many others have indulged in before me, and in which many people believe. I do not think we have sufficient evidence for belief: but I do propose it as a potentially useful concept. If we can manage without it, so much the better, for it is poor science to premise, without evidence, anything that can be dispensed with. But if the case should arise that no better explanation is available to account for the facts, then I suggest that the notion of a universal image-bank is a logical, convenient and useful component for our answer to the question, How does the entity-sighting experience occur?

3.3 THE PSI-SUBSTANCE HYPOTHESIS

In the summer of 1901 two English ladies, Miss Moberly and Miss Jourdain, had an experience in the grounds of the palace of Versailles that has become one of the classic anomaly cases of all time. [52], [104] It has also been one of the most controversial; but there is good reason to think that the percipients were right in believing that they had somehow shared a vision of the gardens as they had been in the eighteenth century, possibly at the outbreak of the French Revolution. Not only did they see the grounds, not as they were in 1901, but as research showed them to have been at the earlier period, but they also encountered several figures whom it seemed possible to identify with persons of that time.

In short, despite some discrepancies (which confuse rather than contradict their interpretation), everything points to the likelihood that the two ladies became involved in some kind of time-displacement. But who, or what, was displaced? Were the two ladies transported back to the 1780s, like time travellers in a science fiction story? Or was the past brought forward to 1901, as when an old newsreel is brought onto today's television screens?

Those who accept the ladies' account at all tend to go for some kind of explanation of the second kind. Time displacement is easier to imagine that way round; besides, there are analogies with ghosts and hauntings, and, more obviously, with normal memory. But even so, the implications are formidable. At the very least, it reinforces the necessity for hypothesizing something along the lines of the 'image-bank' discussed in the previous chapter. For somehow, and (in the loosest possible sense) also *somewhere*, the 'memory' of the 1780s scene had to be preserved, and made available to whoever or whatever chose to repeat it in 1901. Whether we hypothesize, as the two percipients did, that they had 'entered into an act of memory', seeing the place and events as Marie Antoinette had seen them a century or so earlier, or whether we think of them as involved in a less subjective experience, the necessity remains for the record to have been preserved by some means.

For it is not enough to label such experiences 'psychic' and suppose that this absolves us from the necessity of establishing *what* is happening and *how*. As our entire inquiry demonstrates, a psychic experience is as 'real' as any other. Even if we suppose that the picture of eighteenth-century Versailles was given to Miss Moberly and Miss Jourdain in the form of a mental communication, we have not dispensed with the necessity for determining the process by which that communication was effected. The brain is an electrical device; the senses are physical instruments. Beethoven's Choral Symphony was a series of electrical impulses even before it left the composer's mind, long before it was performed by the Vienna Philharmonic and returned to a further series of electrical impulses so that I can play it on my cassette player.

Unless we are prepared to avoid the implications — and as inhabitants of a universe that is presented to us in physical terms, for which purpose we have been equipped with a range of physical senses, I do not think we have the right to avoid them — we have to accept that what happened to the ladies of Versailles was at least partially a physical event. It may be that they picked up some form of psychic signal, from Marie Antoinette or whoever — yet *it has to have been a signal,* like the one out there in the sky that our television aerials catch hold of and convert into electrical impulses, which are in turn translated on our screens into the latest happenings in a soap opera. Alternatively, it may be that they unwittingly stepped inside the frame of a three-dimensional repeat performance of past events, in which case something like a hologram may be a closer analogy — but a hologram, too, has physical reality for all its insubstantiality.

Who arranged for these transmissions to take place, and why, and to what extent they were directed at the two ladies, are other questions which must be asked; but here and now we have to consider the question of how the information, given that it had been preserved in the files of our hypothesized 'image-bank', is made available, what means exist, what process could be employed.

Back in the 1930s, some leading members of the Society for Psychical Research, in London, were considering ideas on the subject of a psychic ether, a non-physical substance on which the events of the world were recorded. It was felt that some such process was demanded by the occurrence of apparitions and especially of hauntings. The Versailles case was one of those specifically mentioned as seeming to require some such hypothesis.

In the F. W. H. Myers Lecture, given to the Society in 1937, C. A. Mace acknowledged the desirability of some such concept, but shied away from any expression of it as crude as a 'psychic ether'.

> Personally I am of the opinion that we can, with a good scientific conscience, postulate the existence of a medium which records impressions of all sorts of patterns of events, and which later or elsewhere may produce a corresponding pattern. We need not ask: what is the intrinsic constitution of this medium: we need not yet ask how it does it. But, however non-committal we may choose to be, we are bound to ask: Under what conditions

do these events occur? Under what conditions does this medium receive impressions and under what conditions will these impressions be revived? [98]

He recognized that there would have to be limits and controls:

> It may be entertaining to think of Nature as engaged in writing a very long book, not a single word of which will ever be read, but such a hypothesis, however entertaining, has no scientific function.

In other words, built into any such concept must be the notion of eventual *use*: our image-bank, like any other, has no meaning unless we also presuppose customers who will make use of its services. And that use would need to be *specific:* total simultaneous playback on a continuous basis of all the past events of world history would be total chaos. But the moment we start to think of limits and controls, we have also to ask: control by *whom?*

In his Presidential Address to the same society two years later, in 1939, the Society's President, the philosopher H. H. Price, took up Mace's tentative suggestions and acknowledged the logical requirement for some such hypothesis. Price's thinking, however, ran towards a 'psychic atmosphere', which would possess 'some of the properties of matter — namely spatial extension and location in physical space — and some of the properties of mind'. [120] He postulated an 'ether of images', which he saw as 'a special and limited form of the hypothesis of a Collective Unconscious'.

But Price was at pains to demonstrate that his 'ether of images' would possess not only the mental properties, which any component of the collective unconscious would have to have, but also material ones: 'Naturally they will not be the same properties as we ascribe to ordinary matter, but they might be somewhat like them.' By way of demonstration, he imagined a haunted room in which percipients had experiences to suggest that somehow traces from the past were persisting there.

> If they are indeed *physical* traces, they must consist in some more or less permanent mode of arrangement of the molecules or atoms or infra-atomic particles, of which the walls, furniture, etc. are composed. And in that case, it ought to be possible to verify their existence by the ordinary methods of Physical Science — by physical or chemical tests of some sort or other. But so far as we know, this cannot be done. It is therefore natural to suggest that the seat of these traces is something which is not material in the ordinary sense, but somehow interpenetrates the walls or furniture or whatever it may be: something which is like matter in being extended, and yet like mind in that it retains in itself the residua of past experiences. And this is just what the Psychic Ether is supposed to be.

He then considered the implications of supposing that the images were derived from the mind of a former inhabitant of or visitor to the house:

Let us suppose that the haunting is of a fairly complex sort, for instance, that the phantasm is seen in a number of different rooms in the house, and is seen to move from one room to another, so that the phenomena are 'cinematographic' rather than just 'photographic'. Here then there is a *group* of persisting images, interrelated in a fairly complex way. Now since the original author of these images is dead, Anti-Survivalists will of course wish to maintain that his mind has ceased to exist. But can they quite maintain this, if our explanation is the correct one? For, to put it crudely, a *bit* of him does still survive, even though his body has long since disintegrated. This set of interrelated images is something like a very rudimentary secondary personality. It was split off from his main personality at the time when he lived in this room; it escaped from his control and acquired an independent existence of its own. And it has succeeded in 'surviving' the disintegration of his body, even if we say that his main personality has not. To be sure, it need not survive for ever. Eventually the images may lose their telepathic charge and fade away. The fact remains that a 'bit' of the deceased personality has succeeded in surviving.

The concept of secondary personality is one which we shall find is of crucial help for our inquiry: it is significant that, in this context, Price also found it a useful concept. His claim, that the phenomenon of haunting necessitates the hypothesis of a trace which has some kind of material dimension, applies no less to our entities. *Wherever they originate, by whatever means they penetrate to the mind of the percipient, those entities are to some degree material artefacts.*

The notion of a 'psychic ether' is just one of the forms proposed for this material aspect. Some are ready to propose a more specific substance, the *astral body*. They suggest that we all have such a secondary body — some even hypothesize two or more — and claim that it is capable of separation from our physical body, given the appropriate circumstances. This, they say, is the basis of out-of-the-body experiences, and also of the kind of voluntary projection discussed in Chapter 2.4. There have been some claims to have proved its physical existence,[3],[80] but these have been ambiguous in replication and cannot be said to be scientifically valid. There is also, however, a great deal of evidence at the anecdotal level of people — not necessarily declared 'psychics' — who 'see' an 'aura' surrounding people they meet, which provides useful indications as to their character, state of health, and so on.

If some such scientifically testable constituent could be shown to exist many of our questions would be answered; but the evidence remains uncertain and we have no choice but to stay with such vague concepts as 'psychic ether'. The idea has been updated in the 'psi-substance' proposed by John Vyvyan[163]: I do not know that he has done more than give it a more acceptable name, but let us be grateful even for that. For if the idea of a substance, part-material and part-mental, is to gain any scrap of support, it needs all the help it can get.

But for those who study these questions, the issue is not whether we can support such a notion, for it is evident that the evidence *requires* the hypothesizing of something as radical and scientifically unacceptable as this: if not a psi-substance, then it will have to be something else equally revolutionary. For somehow we have to account for these entities which appear as a ball of light and slowly grow into full forms; for apparitions that gradually take shape in an empty room; for figures seen by two or more people simultaneously, or by one witness when it leaves the room and by another when it enters another; for entities who bring information, or carry it in the form of identifiable clothing and the like. None of these things can happen without some kind of material dimension; and for that dimension, 'psi-substance' is as good a working label as any other.

3.4 THE INDUCED-DREAM HYPOTHESIS

One of the many puzzling aspects of entity sightings is their ambiguous character, the way they combine elements that seem to indicate an external source with others that seem to refer back to the percipient himself. Religious visions provide us with several examples; thus at Garabandal in 1961 a vision of the Virgin was allegedly seen by a group of children together, which seems to imply a degree of objectivity, while others who were also present could not see her, which suggests a subjective experience. Any attempt to formulate a model that supposes entity sightings to be either wholly external/objective, or wholly internal/subjective, will be inadequate to explain cases like Garabandal.

We have already come across some attempts to resolve this paradox: Edward Gardner's theosophical account of fairy sightings (page 163) was one example. More recently, a number of independent researchers have explored the possibilities of an explanation along the lines of an *induced dream*, drawing on new discoveries about the way the brain works. As it happens, all these studies were inspired by the UFO phenomenon, but there is no suggestion that the processes they hypothesize need be restricted to UFO and UFO-related-entity percipients. Their thinking holds good — or fails to do so — for a wide range of entity-seeing experiences, and one of the researchers whose work we shall consider, Claude Rifat, has in fact abandoned specifically UFO-related research for the more general application of his ideas.

It was precisely the ambiguous character of many UFO reports that, in 1976, stimulated Pierre Guérin, one of France's most senior ufologists, to formulate a prototype 'induced dream' model:

The conclusions I propose are as follows:

1. A notable fraction of the details seen or heard by nearby witnesses of UFOs are an illusion, as if they had experienced, at the time of the incident, a true waking dream.

2. This waking dream does not, as has often been supposed, have its origin in the witness's spontaneous response, for example, his fear when confronted by a UFO. For the emotions felt by a healthy-minded person in consequence of a traumatic event (an accident or whatever) do not, in the ordinary course of things, suffice to create a waking dream of the completeness and intensity of the kind we are here concerned with.

3. On the contrary, everything occurs as though the waking dream was created by the UFO (or by the Intelligence which controls it) according to a predetermined programme, and by means of a technology acting on the mind of the witness in a manner which, admittedly, we do not know, but which could be similar to hypnotic suggestion. In short, and with the proviso that evidence may be forthcoming to the contrary, I think there is no other way of explaining one of the essential aspects of the very many coherent accounts of close encounters reported within a specified period. For this coherence consists not only in the similarity of certain details which seem to reflect the individual or collective unconscious of the witnesses, but also in the concordance between other details which have no affective or cultural content, which could in no way have pre-existed in the witness's mind, and which therefore necessarily seem to have been introduced from the exterior. Thus, the waking dream artificially induced by the UFO would make use simultaneously of the personal elements drawn from the unconscious mind of the percipient, and of other elements imposed by the phenomenon. These latter would include everything which could not have pre-existed in the mind of the witness, such as details of the UFO's structure, the morphology of its occupants etc., and likewise certain absurd physical details, such as the spontaneous re-starting of vehicles which seemed to have been halted by the UFO. [59]

Guérin surmised that by this means the alien intelligences would secure total acceptance of the 'fact' of the UFO, and in the context of this belief they could communicate a message which, under the circumstances, the percipient would have no choice but to accept just as he had accepted the rest of the experience. They would, perhaps, leave him with some such thought as this:

The humanoids who have just emerged from this anti-gravity UFO are extraterrestrials who have travelled from a distant planet; they are good, they know the secret of curing cancer, they require that Earth's inhabitants love one another.

This message would thereafter have the impact of a revelation which the percipient would believe had been addressed specifically to himself. They might add the idea — or he himself might supply it by way of unconscious rationalization — that he had been specifically chosen by the alien intelligences as their envoy on Earth, with a specific mission to his fellow creatures to spread the message. All of which could, of course, equally well be true or false: the aliens could genuinely be motivated by a desire to help us straying Earthfolk into a better course of life, or it could be a sinister preparation for an act of aggression! Though Guérin does not make the point, it is obvious that the same model would serve very well for other categories of entity besides extraterrestrial aliens. Those who believe that some visions are of deceitful spirits could well adopt such a hypothesis, which would explain the puzzling ambiguities in such cases: it could also be the basis for the no less bothersome encounters with alleged spirits of the dead.

A year after Guérin published his thesis, the ufological world was shaken by the 'defection' of one of France's most respected investigators, Michel Monnerie. His encounters with UFO witnesses convinced him that they were deluded; at the same time he could not deny their patent sincerity. To account for this paradox he, like Guérin, hypothesized that they were being subjected to a waking dream. The difference was that in his case, since he did not believe there were any physical UFOs or UFO-entities to create the dream, he concluded that it was a wholly subjective experience. So for Monnerie the process of seeing a UFO — which we can extend not only to include UFO-entities but other kinds of apparition — had no external agent; at the most there might be an external stimulus to trigger off the percipient's mind. But from then on, all the creative work was carried out within his own mind, a psychological process involving a projection of elements of personal or cultural significance, moulded into a coherent form by the ever-inventive 'dream producer', whose existence we ourselves have had to hypothesize at the outset of this inquiry. [105] [106]

Clearly there is a crucial difference between Guérin's and Monnerie's thinking, hinging on the question whether the elements forming the reported experience could reasonably be supposed to be latent in the percipient's mind: Monnerie thinks they could, Guérin disagrees. We may note, in passing, that something of the same dilemma is found in the DeHerrera/Lawson imaginary abduction experiments described in Chapter 2.1; there, too, we have to ask ourselves if a ufologically illiterate subject could have found, within the resources of his own mind and memory, the necessary construction materials for the imaginary encounter he so fluently narrates, or whether we have to suppose some kind of external data source.

Claude Rifat, whose work in the same field carries the hypothesis an important step further forward, recognizes this difference of interpretation but points out that his model could be valid in either case. Rifat is a Swiss biologist, whose knowledge of the brain enables him to take the induced-dream hypothesis a stage further by showing by what kind of physiological process the dream could come about.

He takes as his premise that the reports we receive of UFO encounters are imaginary descriptions, even though they may contain some real elements; though they purport to relate to external events, and are sincerely so related by the witness, they contain much material or coloration which must be seen as related to the percipient himself. As Rifat points out, the same admixture is found in dreams, where all kinds of objective elements are presented, yet in an emotional context and with a 'directedness' that is pertinent to the dreamer and no one else. Consequently, he argues that we need to find the physiological mechanism in the human brain that could be involved in the process of moving from the waking state, in which exogenous (= coming from outside) information is received, to the sleeping state which makes use of endogenous (= originating inside) material. (At the same time he reminds us that there are certain intermediate states, such as those induced by drugs, in which we receive exogenous sense impressions but distort them subjectively.)

In the original presentation of his theory[131] Rifat pointed to a specific portion of the brain — the *locus coeruleus* — as being indicated as the switchgear required. As I understand it, it is this component which is for the brain the equivalent of the switch on my hi-fi that determines whether I hear incoming radio signals or home-produced material from my tape library.

More recently[132] he has modified his theory in the light of the latest neurological research, which shows that dreams are associated with the *'Raphé'* system, a serotoninergic system that interacts with the *locus coeruleus*. It is the modulation of this system that releases hallucinations, dreams, delirious visions and so forth. The details, for our purpose, are unimportant: what *is* important is that there is a physical/chemical basis for the process which, while it does not control the *content* of the resulting experience, does to a large extent determine what *kind* of material will go to make up that content.

Rifat reminds us of the fact that micro-waves can affect the functioning of nerve cells: from which it is reasonable to argue that if the percipient were exposed to an emission of micro-waves, this could have the effect of initiating the kind of *process* we are thinking about, leading to the kind of *experience* we are thinking about. Where would such micro-wave radiation come from? Rifat takes no sides on this issue, leaving us to conjecture whether they are deliberate or accidental. However, continuing to hypothesize for the time being in terms of UFO entities, we could face a choice of options:

★ An effect deliberately induced by the aliens (who may be the extraterrestrials they seem to be, benevolent or otherwise, or masquerading spirits, or whatever).

★ A purely accidental effect: the propulsion system of UFOs, say, *happens* to give off these micro-waves, and they *happen* to have this effect on us.

★ An effect that is accidental, so far as the UFO is concerned, but is in part voluntary on the percipient's part, as his unconscious mind, stimulated by purely physiological factors to fabricate an induced dream, proceeds to weave a coherent and meaningful narrative that will satisfactorily account for the experience in terms of his own beliefs, hopes, fears and so on.

Inevitably, because Guérin, Monnerie and Rifat were all primarily concerned with the UFO witness, their hypotheses are presented in terms of UFOs and UFO-related entities. But it is evident that their ideas are susceptible of a wider frame of reference, and if something like this is happening, then it is a kind of process that should be of the greatest interest to psychologists and sociologists alike. For here we have an area in which purely physiological factors, which can be expressed in terms of chemical substances and physical mechanisms, are interacting with processes which, whether we designate them as 'creative' or 'psychic' or whatever, are *not* reducible to biological and chemical terms. What triggers the experience may be the action of a specific kind of radiation on a specific component in the human body: but the selectivity and creativity that weave together the random elements thus summoned forth is a process on an altogether different level. It could be precisely an interaction of this sort which is an essential component in our model of the entity-experience.

3.5   THE DISSOCIATED
PERSONALITY

The concept of multiple personality has cropped up in several contexts already during our study. Of all known psychological processes, it is probably the one that promises to throw most light on our problem. Its importance for our study is the hints it offers as to *what* may be happening in an entity-sighting experience, and *how* it happens.

Multiple personality is a special form of dissociation, in which the subject has, or gives, the impression that he contains more than one person. These persons have names, distinctive personalities and consistent behaviour patterns and value systems, usually in sharp contrast one from another: they 'take it in turns' to control the subject, causing him to seem to others to shift — often abruptly — from being one kind of person to being another. There is nearly always a degree of conflict between the personalities, which can be carried to extreme situations in which one actually wishes to kill another.

This kind of disintegration is, of course, ultimately harmful to the subject, and is an ailment doctors undertake to treat. Initially, however, it occurs as a short-term solution to a state of affairs or situation in the subject's life with which he cannot cope, so that there are real benefits for the subject. The doctor's aim, therefore, is to persuade the subject that there are better ways of solving the difficulty, ways which do not involve this extreme solution. Having done that, the doctor can then persuade the secondary personalities that they have no further purpose, whereupon they can be compelled or persuaded to disappear.

The situation that leads to the creation of secondary personalities is nearly always some kind of childhood crisis, generally related to sexual assault, parental neglect, violence from other children, and so on: almost without exception, cases reveal a history of extremely disturbed childhood. The incidents themselves may be of a kind with which other children manage to cope successfully, but with which this particular subject, for one reason or another, cannot come to terms by normal means. A timid American coloured boy named Jonah, for example, developed a 'warrior' personality who took over when he was attacked by white children:[70] a girl whose father made no secret of the fact that he had been hoping for a son, developed a 'boy' personality, which sought to give him what he wanted.

Though little understood, multiple personality is now widely recognized as a specific psychological disorder, and there is a growing literature on the subject. Alternative interpretations of the facts have been made, not only by observers but also by the subjects themselves, who have claimed that what is taking place is not a disintegration of a single personality, but an 'invasion' from outside by spirits who 'possess' the subject on a temporary or intermittent basis. It must be said that some of the evidence for this explanation is very persuasive, and at least one highly experienced doctor specializing in multiple personality cases has admitted that, with the greatest reluctance, he is forced to conclude that some of the personalities he has encountered in his subjects are indeed invading entities or spirits, and not simply different facets of the subject's own being. [2] Fortunately, whether this is so or not is not of direct relevance to our inquiry, so we can leave the question unresolved.

Another claim that has been made is that the specific disintegration into separate personalities is an iatrogenic condition — that is, one caused by the doctor himself, imposing his own interpretation on the case. Admittedly there is some evidence that in the course of treatment reinforcement and modification take place; it is of course necessary for the doctor to identify the various personalities in order to recognize the part they are playing in his patient's life, and the mere act of identification could have the effect of reinforcing their identity and strengthening their role. But this is not the same as creation, and all the evidence seems to point to the view that the division has occurred before the case comes to the doctor's attention.

Nevertheless, this suggestion reminds us, on the one hand of hypnotic subjects' readiness to go along with what is suggested to them, which could find a parallel in a disordered person creating a secondary personality to 'please the doctor' by presenting him with, as it were, an incarnation of the patient's troubles; and on the other, of the ability of people to fabricate at a moment's notice a highly organized fiction, such as the age-regression subject we noted who was able immediately to find a role for the twentieth-century hypnotist to play in his past-life fantasy scenario. In creating his secondary personalities, the multiple personality patient may be using the same remarkable skills as in some of our entity cases. We should certainly explore the possibility that dissociated personalities represent a different kind of application of the same basic processes: even that *the conditions which in one subject lead to multiple personality states have much in common with those which in another subject lead to entity fabrication.*

Most of us, most of the time, suit our behaviour to our company: as householders we may be timid but as business executives aggressive; in a club with colleagues we may be patient and considerate, while as motorists in the street we are quite the contrary. Most of these shifts of attitude are unconscious, and for the most part are not so extreme as to cause more than a passing comment that so-and-so is 'quite a different person' when he is away from his parents or his wife. In a sense, multiple personality cases are simply more structured examples of the same process; but because the situation is more demanding, stronger measures have had to be taken. The subject cannot find within his normal personality the characteristics he needs — the behaviour pattern required to deal with the situation is just not compatible with his everyday self. So his subconscious mind 'exteriorizes' the situation by fabricating a personality who is so different from his normal self that the question of incompatibility does not arise — and so his second self can go ahead and do things which, in his normal self, he could not — or believes he could not.

But there are other courses open, and one of those could be to exteriorize yet further, and 'create' an entity who does not purport to be another personality of the subject, but someone outside — either someone known to the subject, or a stranger. And so we might get Bernadette Soubirous believing that she has been visited by the Virgin Mary, and a UFO contactee believing he has been approached by a friendly extraterrestrial, each bringing to the subject the comfort, reassurance, ego-supporting confidence or whatever other benefits the subject lacks.

Some support for this proposition comes from the best-known and most richly-documented of all multiple personality cases, that of Miss Christine Beauchamp, who was from 1898 to 1904 in the care of Dr Morton Prince of Boston. [121] The case was in many respects typical: the 23-year-old Christine had three main personalities, which alternated their 'occupancy' of her physical body. With the central aspects of her case we are not concerned, as there is no direct relevance to entity studies: but some of the marginal phenomena associated with her case are extremely suggestive. First, there is the parallel with religious visionaries of the kind we considered in Chapter 1.8:

Miss Beauchamp as a child frequently had visions of the Madonna and Christ, and used to believe that she had actually seen them. It was her custom when in trouble, if it was only a matter of her school lessons, or something that she had lost, to resort to prayer. Then she would be apt to have a vision of Christ. The vision never spoke, but sometimes made signs to her, and the expression of his face made her feel that all was well. After the vision passed, she felt that her difficulties were removed, and if it was a bothersome lesson which she had been unable to understand, it all became intelligible at once. Or, if it was something that she had lost, she at once went to the spot where it was. On one occasion when she had lost a key, her vision of Christ led her down the street into a field where under a tree she found the key. She constantly used to have the sense of the presence of some one (Christ, or the Madonna, or a Saint) near her, and on the occasion of the visions it seemed simply that this person had become visible.

On the night of the very day when the account of her early visions was given me by B II [her second personality] and confirmed by the Real Miss Beauchamp, the latter had a vision of Christ which I was able to investigate. Miss Beauchamp had lost a bank cheque, and was much troubled concerning it. For five days she had made an unsuccessful hunt for it, systematically going through everything in her room. She remembered distinctly placing the cheque between the leaves of a book, when someone knocked at her door, and this was the last she saw of the cheque . . . That night she was unable to sleep, and rose several times to make a further hunt. Finally, at 3 o'clock in the morning, she went to bed and fell asleep. At 4 o'clock she awoke with the consciousness of a presence in the room. She arose and in a moment saw a vision of Christ, who did not speak but smiled. She at once felt as she used to, that everything was well, and that the vision foretold that she should find the cheque. The figure retreated toward the bureau: she then walked automatically to the bureau, opened the top drawer, took out some stuff upon which she had been sewing, unfolded it, and there was the cheque along with one or two other papers.

Neither Miss Beauchamp nor B II has any memory of any specific thought which directed her to open the drawer and take out her sewing, nor of any conscious idea that the cheque was there. Rather, she did it, so far as her consciousness goes, automatically. B II, however, was able to give facts which make the matter intelligible. Miss Beauchamp remembers distinctly putting the cheque in a book, but B II says that she did not actually do this. She held the book in one hand and the cheque in the other with the full *intention* of placing the cheque in the book, but at this moment a knock came at the door. Thereupon she laid the book and the cheque upon the table. After answering the summons she went to the table and picked up her sewing and unconsciously at the same time gathered up the cheque . . .

It is pretty clear that the finding of the cheque was accomplished automatically by a subconscious memory, and that the vision of Christ was the resuscitation of an old automatism, under the influence partly of this subconscious memory, and partly of the suggestion derived from our conversation a few hours previously.

And he concludes:

Visions like those of Christ and the Madonna, which express the conscious or the subconscious thought of the individual, are from the point of view of abnormal psychology to be interpreted as sensory automatisms of which the genetic factor is the person's own consciousness [my emphasis].

Whether we can conclude that the entire vast case material of religious visions should be interpreted in this way is a question to which we shall return in our concluding chapter. But the significance of such an incident occurring in the case history of a multiple personality subject is in itself remarkable, and confirms the probability that the two kinds of experience have much in common.

The case of Christine Beauchamp also contains another clue which I believe to be of fundamental significance: indeed it was my coming across this clue which enabled me to see how this whole assemblage of disparate material could be given some kind of structure and coherence.

This particular incident occurred at a time when 'a contest between personalities' was taking place between B IV and a somewhat feckless personality called 'Sally'. One day, B IV had the following unnerving experience:

> While I was brushing my hair a sensation of great fatigue came over me, the effect of the exciting day I had just passed through. I finished, however, and then as I sat in my chair, I stooped to change my slippers when, with a sudden shock of horror, I saw directly facing me at the opposite side of my room my own feet. They were white and shining against the black background. I fell back in my chair overcome.
>
> At once I was conscious of pain in my legs below the knees and of a feeling that my feet were gone. I felt for the moment certain that this was the fact, for I had no sensation below the seat of the pain. My legs seemed to end in stumps and I instinctively leaned forward to protect them with my hands, keeping my eyes fixed upon the feet opposite to me. But the next moment I realized that this was but another device of Sally's, to prevent my going out. I told myself this over and over again. 'It is only Sally,' I said; yet I could not move or take my eyes from those feet . . . Sally had always treated my body as if it were not even remotely connected with herself, cutting, scratching, and bruising it in a way so shocking that it is hard to believe. Could she now have gone farther and really have done this? It did not seem impossible.
>
> I was in the greatest pain and could feel nothing below my knees. Finally, making a great effort, I threw myself on the floor and dragged my body across the room. I brought myself near enough to touch the feet; — they were bloody. Making a supreme effort to touch, ever so lightly, the nearer one, I found my fingers stained with blood.

B IV was of course quite right in thinking that this was Sally's doing, and Sally confirmed as much to Dr Prince, telling him, 'I can make her not see things, or see things.'

> Sally subconsciously induced in B IV and B I time and again hallucinations which were visual representations of her own subconscious thoughts . . . *The evidence is conclusive that subconscious ideas can excite hallucinations in the primary consciousness.* It follows that we may not be able to determine the genesis or origin of any given hallucination without knowing the content of the subconsciousness. If someone versed in abnormal psychology had hypnotized the numerous saints and sinners who have experienced visions and voices and examined their individual consciousness, we should know much more about the origin of their hallucinations.

The significance for this study is, I hope, self-evident. Here we have evidence of a capability inside the mind for generating hallucinations, in a way which must be very gratifying to the champions of the 'all in the mind' interpretation of these things. But we also see that it is not quite so simple; for, in another sense, this is *not* the internal process it appears to be. For, so far as 'Sally' was concerned, B IV, her victim, was a separate person: that is to say, the induction by one part of the mind of a hallucination intended for another part of the mind has something of the quality of an exteriorization. From which it follows that *one of the distinguishing factors which has bothered us throughout this study, the question of the degree to which entity-sightings are internal or external in their origin, is seen not to be so rigidly watertight a distinction after all.*

It would be a mistake to attach too much weight to this piece of evidence without taking other considerations into account: multiple personality cases are, after all, extreme psychological situations. Nevertheless, as in so many other fields of research, extreme cases can highlight a feature that is not easily distinguishable in the normal course of things. Here we have evidence that there exists, within the human mind, a means whereby one part of the mind can make another part experience the illusion of seeing an entity, in such a way that the conscious mind accepts it as a real and objective experience, giving no indication of whether its origin is internal or external. The simple fact that such a means exists entitles us to ask whether it may be operative in the phenomena we are trying to explain.

We may also speculate whether there are any limits to the kind of hallucination which can be manufactured in this way? How does the secondary personality relate to that part of us that goes travelling in out-of-the-body experiences, that appears to acquaintances at the moment of our death, or to the replica of ourselves that stays on Earth to haunt when the rest of us moves on to whatever awaits us in another life?

3.6 HALLUCINATIONS, PATHOLOGICAL AND OTHERWISE

In 1901 a doctor, Henry Head, presented the annual series of *Goulstonian Lectures* with the unremarkable title of *Certain mental changes that accompany visceral disease.* [67] Among the side-effects he noted was the causing of hallucinations in patients, and these he analyzed in some detail in the course of his study.

Seven years previously, the Society for Psychical Research had conducted its celebrated Census of Hallucinations, [148f] whose organizers had, as far as possible, excluded all cases of pathological origin. So now a member of the Society's Council, J. G. Piddington, recognized that Head's study provided a unique opportunity to make a comparison between the two kinds of hallucination, those of pathological origin and those which were not, which would establish whether there were any differences between, on the one hand, those experiences activated by physical causes and, on the other, those caused by mental or psychical stimuli.

Piddington's paper [118] occupies sixty pages of the SPR *Proceedings*, but a few examples of his comparisons will be sufficient to indicate his findings:

	PATHOLOGICAL CASES [from Head study]	PSYCHICAL OR MENTAL CASES [from SPR Census]
1 Circumstances of experiencing the phenomenon		
Waking from sleep:	Usually	12 per cent (38 per cent lying awake in bed)
Lighting:	dark or dusk	no limitation
2 Appearance of entity seen in hallucination		
Figure perceived:	65 per cent of all cases	75 per cent
Clothing:	Shrouded or draped	everyday clothes
Face:	misty	identifiable
Bearded, if male:	never	normally (this was normal for 1894)
Sex:	doubtful	certain

	PATHOLOGICAL CASES [from Head study]	PSYCHICAL OR MENTAL CASES [from SPR Census]
Legs:	not generally seen	virtually always seen
Overall appearance:	corpselike	natural

3 Behaviour of the entity

Movement:	usually stationary, sometimes some slight movement	natural
Sound:	never	16 per cent of cases
Gestures, changes of expression:	virtually never	commonly
Colour:	never	very common
Disappearance:	unrealistic	realistic

4 Effect on witness

Depressing:	generally	so far as is known, never
Frightening:	almost always	hardly ever

5 Overall character of the phenomenon

Repeated or recurrent:	frequently	exceptionally
Multi-sensory:	uncommon	common
Veridical:	none reported	common
Collective (seen by more than one percipient)	never	occasionally
Localized in space:	seemingly never	frequently
Includes touch:	seemingly never	11 per cent of cases

On the strength of these and other contrasts, Piddington felt justified in concluding: 'I do not believe that there is a single case of hallucination printed or referred to in the Census Report which completely falls into line with the visceral type', and he claimed that these figures confirmed the conclusion which the compilers of the SPR Census, although Head's study had not then been published, had reached: that they had isolated a phenomenon which has a genuinely psychical origin, as distinct from the classic pathologically-induced phenomenon.

Insofar as they had identified a distinction between the two categories, they were unquestionably justified in their belief. We may reasonably suppose that the physical mechanism for the generation of hallucinations is the same in both cases, but it is no less evident that some factor must have entered into the matter, to cause so many clear-cut differences.

But what is that factor? It does not seem that we are here concerned with the kind of external agent that might be involved in a ghost or a haunting case: rather it seems more likely that the origin is to be found in the percipient's mind. But that does not rule out the probability that the phenomenon had an external *stimulus,* perhaps some kind of extra-sensory perception.

In crude terms, we can say that the experience, in both the pathological and non-pathological cases, consists of three parts: stimulus, process, and phenomenon.

As regards the *phenomenon* itself, the end-result as it is perceived by the witness, it would be an assumption that we have no grounds for making to suggest that there is a difference in *kind* between the two sets of cases. All the differences relate to the *character* of the hallucinated entity and the clarity and verisimilitude with which it is perceived, and we may reasonably attribute these to a difference in the initial stimulus.

There seems no reason to doubt that the *process* is the same in both classes. But whereas in one class, the *stimulus* arises as some kind of side-effect of illness, with no specific incident to, as it were, press the button that sets the machinery running, in the cases reported by the SPR Census the experience can in many cases be related to some such incident as an accident to or the death of someone known to the percipient. This stimulus, because it is more specific, activates a more highly developed kind of hallucination — seen more clearly and in greater detail. Also, because in this class of cases the emotions are involved, the hallucination is more vividly experienced, and this seems a reasonable explanation for the greater degree of verisimilitude.

Many of the contrasting details can be accounted for by reference to the origin of the stimulus. An accident can happen at any time of day or night, so the non-pathological hallucinations are reported as occurring at any time: the pathological instances, on the other hand, seem to occur only when conditions are favourable to sensory illusion, or possibly when the physical state of the patient is suitable.

We may reasonably suppose that in the case of the pathological hallucinations, all the material is derived from the percipient's own 'store', whereas in those that have an external origin, at least some of the material will be contributed by the external stimulus, thus accounting for the greater substantiality of such cases — the natural clothing as opposed to the vague drapes and shrouds, the identifiable features and so on.

All in all, Piddington's findings present us with a solid demonstration of an entity phenomenon whose origin can be internal or external; where the fundamental character of the phenomenon is in all probability the same in either case, but where the difference in the originating stimulus leads to corresponding differences in the eventual experience. It will be interesting to see if what is true of hallucinations is true also of other kinds of entity phenomena.

3.7 THE SCIENCE FICTION PARADOX

In November 1964 a 29-year-old American housewife, referred to under the pseudonym Mrs Merryweather, described how, over a period of four hours, she and her mother-in-law had watched the crews of two grounded UFOs, perhaps a dozen men in all, carrying out repairs on one of their craft. [44b] She seemed to be a reliable witness, and told her story coherently and convincingly, methodically yet with traces of humour. As an account it held together well; it was confirmed to some extent by other witnesses, and traces were found at the location she indicated. Leaving aside for the moment the inherent improbability of the event, what reasons have we for being reluctant to believe her story?

There is really only one factor that in any way speaks against the report, and that is the witness's personal psychological make-up. There were stresses in her family situation, investigators found, and these seem to have triggered off some psychic happenings; these were accompanied, as happens frequently in America, where religion is so live an issue, by a personal religious uncertainty which a year after the incident led to Mrs Merryweather becoming a Mormon.

If the incident was a purely factual matter, there should be no connection between the percipient's religious misgivings and the thing she and her mother-in-law reported seeing. Apart from the fact that she used binoculars to observe the repair crews more closely, the event seems to have been as lifelike as though it was a car or a helicopter which was being repaired. The issue seems simple enough: either it happened or it did not.

It ought to be that simple. What stops it being simple is another simple fact, that the crew managed to repair the ship. French ufologist Bertrand Méheust has commented on the incident: 'Can the reader picture the scene, as the dawn rises over New Berlin next morning, and sees the crew with oil on their faces, frustrated in their attempts to repair the machine, and captured — or clobbered — by American military forces while an excited crowd watches?' Yes, we can picture it, as we can picture a movie scene: but we do not believe that it could have happened in real life. It did not happen like that in November 1964, it never has happened, though stories of UFOs being repaired by their crews are one of the archetypal patterns of UFO reports. Somehow the UFO crews *always* get their repairs done, under cover of darkness, undisturbed by passers-by except for the witness himself (who is usually alone or almost so, isolated and in no position to summon others); always the aliens manage to get airborne before the first commuters start passing by on their way to work.

Méheust made this point in a book[103] that adds a whole new dimension of the bizarre to a problem which is quite bizarre enough already. Briefly, his thesis is this: there are correspondences between present-day UFO sightings, on the one hand, and early science-fiction stories, on the other, which are so close, and so many in number, as to make coincidence seem out of the question: in one or two cases we might allow coincidence as an explanation, but not when so much detail is entailed, and not in case after case. Instead, we must look for some cause-and-effect sequence.

Méheust presents us with a book full of supportive evidence, showing that encounters reported in the 1970s, by such witnesses as illiterate Argentinian farmers, parallel in close detail stories told as long ago as the 1890s, many of them printed in French and, so far as is known, neither translated nor republished. For example, in 1934 a Belgian writer, Ege Tilms, wrote a pulp fiction story in which a character sees a UFO, approaches it, is drawn to it by a strange force; a door opens, and he finds himself in a small room, with no obvious source of light, and no furniture except a seat and a small table . . . In 1958 a Brazilian farmer described how he had seen a UFO, and had been seized by small beings who took him on board a vessel, where he found himself in a small room, lit by fluorescent light from the ceiling, where there was virtually no furniture . . . And so Méheust confronts us with parallel after parallel, sometimes relating to the UFOs themselves, sometimes to their actions and mode of movement, sometimes to the associated entities and the way they behave, sometimes to the encounter experience itself.

Just to take one specific point which I personally have always found intriguing. On 26 July 1978, 37-year-old 'Gerry Armstrong' was hypnotically regressed to the occasion when, as a twelve-year-old schoolboy in England, he was taken aboard a UFO and given a physical examination by its occupants. His account is more or less a standard one, but from the words he uttered under hypnosis, I note the following: 'In a room . . . ain't no electric light bulb. Can't see a bulb . . . I can't see any bulbs. And then one go through the wall. I didn't . . . can't see a door.' Later, going over the material he had spoken while hypnotized, but now in a wakened state, he said, 'I'm being put in a room . . . I think it's a room. I'm trying to understand where the light is from. It's very, very interesting. We seem to walk through a wall, but there must have been a door.'[62]

In 1977 40-year-old Betty Andreasson was hypnotically regressed to when, ten years earlier, she was taken aboard a UFO and subjected to a number of astonishing experiences during a sequence of events which largely conformed to the standard abduction scenario. From her account I note:

Betty: Whoosh! Another door opened. And you can't even see those doors. They just go up when they open.
Questioner: Can you see the source of the illumination?
Betty: It comes from all over the place.
Questioner: Can you see any welded seams on the wall or some type of seam?
Betty: No, it seems smooth all the way around. [46]

You may wonder why the questioner put those particular questions to Betty. The answer is that a preoccupation with the source of illumination, and with the way in which the abductees enter the UFO, or rooms within it, without seeming to pass through doors, is one of the characteristics of this type of case. But not only that: it is also a preoccupation of science fiction writers, as Méheust shows in his book. He quotes this example from a French science fiction story written in 1908: 'A green light, diffuse, shone on them, but where did it come from? It seemed somehow to emanate from the very material which formed the spherical wall of their strange room . . .', and this from a Belgian story of 1934: 'A pale light enabled them to distinguish their surroundings, even though there was no visible source of light.' Similar examples could be quoted with regard to the seeming absence of doors.

Why this strange preoccupation? One answer might be that lights and doors are things all human homes and vehicles possess, so they are something we can relate to when we find ourselves in an alarming situation. But so are lots of other things which do not crop up in the accounts in this way: why, when something so extraordinary as being taken aboard an alien spacecraft is happening, do such trivial details seem so fascinating both to the allegedly real-life witnesses and to the writers of fiction?

This preoccupation with doors and lights is trivial in itself, but it is precisely for that reason that I draw attention to it; for it is the correspondence of this kind of seemingly irrelevant trivia that confirms the extraordinary parallels between fiction and alleged fact. That the writers should have predicted, long before it was ever reported in real life, that humans would be abducted on board alien spacecraft, was, though sufficiently surprising, not outside the reasonable limits of imaginative speculation. But that they should describe the happenings in such detail, and with the same curious preoccupations, shows that something more than artistic imagination is operating.

We have a choice of possible explanations:

★ that despite the improbability, the present-day UFO witnesses did know of the old science fiction stories, and for reasons of their own chose to pretend that something of the sort had happened to them in real life;

★ that the old science fiction writers had in fact experienced genuine UFO encounters similar to those of the present day; but, the prevailing cultural climate being unfavourable to such happenings, they repressed the memories of their experiences either consciously or subconsciously, but, feeling obliged to tell their story somehow, presented it in the form of fiction;

★ that the old writers, thinking they were writing fiction, were unconsciously having premonitions of the real-life experiences which UFO witnesses would experience a generation later.

None of these is convincing. Méheust himself puts forward the tentative suggestion that both the science fiction writers and the present-day UFO percipients are 'feeding in the same imaginary stream', in the dreams of an epoch in which technology and magic are fused. Speaking specifically about the New Berlin sighting, he suggests that it is typical in its combination of 'ostentation' with 'elusiveness'. The UFOnauts show themselves, put on a display; but they do not go beyond a certain point; they do enough to make the witness, and we who hear her story, wonder and speculate, but not so much as to force us, and the community as a whole, to face up to the fact of the reality of the UFOs and their occupants.

Méheust's findings are most easily accounted for if we hypothesize that somehow both his fiction writers and his 'true' witnesses are obtaining their material from a common source: which brings us back, once again, to the 'image-bank' hypothesis and the notion of a collective unconscious. When Jung wrote his book on flying saucers, he had nothing like the wealth of material available to him that now exists; even so, he was able to see the significance of the first contactee stories. Unquestionably he would have seen, in the classic abduction scenario that has since emerged, ample confirmation of his views. Over and over again that scenario has been reported, in seeming sincerity, by witnesses who have subsequently withstood investigation to the extent that their stories have not been shown to be false, even if it has been equally impossible to prove them true. American author Budd Hopkins has compiled a collection of such cases[73] which present the same detailed characteristics in case after case, where the probability that the percipient could be aware of other such cases seems virtually nil. Hopkins accepts this as evidence that these abduction claims are real:

> What the purposes of these temporary abductions are, and what part of the experience may be purely psychic, we can only guess, but that they have a physical dimension seems to me beyond doubt. Several abductees bear scars on their bodies from incisions made years earlier when the subjects had been children. On separate occasions, I have heard these witnesses, under hypnosis, describe in almost exactly the same words the equipment used to make these incisions. Although these abduction accounts include almost nothing that can be construed as being deliberately hurtful or malevolent

on the part of the abductors, the pattern that emerges, nevertheless, leaves me thoroughly alarmed . . . We have no idea how many such kidnappings may already have taken place, but I believe there are vastly more than the mere two hundred or so incidents which have been investigated.

Hopkins then produces evidence to show that this figure must represent only a very small proportion of the cases which have in fact occurred, and concludes that 'we can logically theorize that there may be tens of thousands of Americans whose encounters have never been revealed', while the number worldwide is almost beyond conjecture.

If these abductions are a physical fact, then the human race is being confronted with just about the most formidable challenge of its entire history: yet even if there is no physical basis, there is a sociological reality at work — Jung's collective unconscious operating on a massive scale.

Once again we are up against the internal/external, subjective/objective dilemma. Let us get back to the New Berlin case once more, and consider the alternatives:

1   We can believe that the incident took place on the physical plane, as reported. This is in many ways the most attractive alternative; after all, not only does the witness present the story as if it were observed fact, but she adds convincing details, she tells us how she watched the repairers over a long period, and through binoculars, and how her mother-in-law saw it too. But it does also require us to accept that her psychic history is irrelevant; that the similarity of her story to other UFO repair stories is simple coincidence; that the lack of physical evidence (imagine spending half the night repairing a car and not leaving a drop of oil, a loose washer or nut, or even a mark or a footprint on the ground) is explicable; and that the UFO happening to break down in such a remote place, in the middle of the night, and away from observation almost entirely, was just a lucky chance.

2   We may suppose that the incident was a display staged for Mrs Merryweather's benefit, by 'real' UFO entities. Improbable as it may sound, this is an interpretation put forward by many ufologists, who suggest that the entire UFO phenomenon is part of a carefully orchestrated programme of cultural conditioning, whereby the UFO people subject us to sightings of UFO incidents of gradually increasing elaboration, so that we do not suffer too great a culture shock when the day comes that full and open contact is made with the extra-terrestrials.

Some support for this scenario is given by Méheust's point about the repairs being completed 'on time'. The whole set-up — the secluded location, the small hours of the morning, the distance from the few observers — all these facts, which make the physical hypothesis improbable, strengthen the possibility that the event was an artefact deliberately contrived.

On the other hand, the suggestion that a whole team of extraterrestrials, and two expensive UFOs with their surely high maintenance costs, would devote several hours to a performance designed to educate two American housewives to the possibility of inter-planetary communication, is something less than plausible.

3   But perhaps the UFOs and their crews were not actually there? Perhaps the whole thing was a fantasy, induced by the aliens with the same motives as in the preceding scenario, but effected by holography or some such advanced technology; so what Mrs Merryweather and her mother-in-law were watching was a clever dramatic presentation. It is possible that Mrs Merryweather's psychic potential is relevant here: perhaps, to see such a presentation, you need to be a special kind of person; Mrs Merryweather may have been such a person, so may her mother-in-law, or the latter may have picked it up from her daughter-in-law by suggestion or extra-sensory communication.

4   Alternatively, perhaps we can dispense with the aliens altogether, and see the incident as an *internally* generated fantasy, composed by Mrs Merryweather's subconscious mind and mounted by the 'dream producer' whose existence we have previously hypothesized. If we accept this scenario, however, while welcoming the opportunity to discount external intervention by aliens or deceiving spirits or whatever, we also have to recognize the necessity of indicating *some* source of inspiration. We may go so far as to suggest that there may have been a real-life event to trigger the supposed manifestation; but from that point, Mrs Merryweather's mind would have to find the material she needed — perhaps a newspaper account she read casually and immediately forgot, perhaps stories heard from acquaintances, but added to something deeper, an archetypal structure which may have had its origin in the collective unconscious and for which the appropriate construction material is available from the image-bank.

Science fiction is a very loosely defined literary genre: just about the only thing that all science fiction has in common is what distinguishes it from other forms of fiction. Other fiction accepts, as the setting for its imagined events, a world that is more or less the world as we know it: science fiction, on the other hand, either explicitly or implicitly, predicts the future or imagines an alternative present (or, rarely, an alternative past). Instead of dealing with an 'is now' it supposes a 'will be' or a 'could be'. This is what we all do, every night, in our dreams, and every day, in our waking fantasies: if the entities seen by our percipients are fantasies, then it is not surprising that there are similarities with those created by science fiction authors. It could be that Mrs Merryweather, and the imaginary abductees, and their supposedly real-life counterparts, and all the others whose experiences seem to straddle the borderline of reality, are, in effect, writing their own science fiction.

3.8 THE GRAFTING PROCESS

There is one category of entity sighting that seems comfortingly straightforward, for it is demonstrably *not* 'real' in any way:

> The woman who reported this described it as a 'star', only much brighter. It was positioned low in the south-west sky, starting around 7 o'clock in the evening on January 30, 1976 — exactly where Venus was located at that time. She did not see Venus in addition to this object. She then watched the light descend gradually to the horizon during an hour's period of time, which is exactly what Venus would do. This setting motion was perceived by her as being 'jerky'; her husband thought that it was only a star, but she encouraged him to perceive the 'jerky' descent too, which got him excited. After staring at it for a sufficiently long time, the woman became convinced that she was looking at the illuminated window of a UFO and that she could see the round heads of the occupants inside, heads with silvery-coloured faces. She then proceeded to see this apparition in the same place every night for successive nights. I told her that it was Venus. Her reply: 'You are talking to a woman fifty-four years old. I know what stars look like!'[68]

The author, Allan Hendry, cites another instance in which a UFO was drawn by children with windows through which occupants could be seen, yet it was definitely identified to be an advertising plane. His comment was that 'the desire to fill in a domed saucer body is a common one in ad plane reports. The presence of little green people looking through three windows is not.' But there was a very similar instance in West Wales during the 1977 UFO sighting 'flap'; a number of schoolchildren, who claimed to have seen a UFO on the ground in a field near to their school, also claimed to have seen UFO-entities in its proximity, and described them in terms of space adventure stories such as *Star Trek*. Investigation showed that the entities could hardly have been seen in such detail at such a distance, that only a few of the children had seen them, and that their accounts of their numbers and appearance were very varied. This did not prevent the children being absolutely terrified by their experience.[37, 124]

It is sufficiently evident that in cases such as these the percipient is filling out the picture with what he feels 'ought' to be present — a process which comes as no surprise to the psychologist, but is apt to catch the unwary UFO investigator off his guard.

But there is more to it than that. Here we have unequivocal evidence that, on some occasions at least, percipients are claiming to see extraterrestrial entities who can be shown, without question, *not* to be there. Could the same process be operating in other categories of entity sighting? And if so, what is the process?

We have already noted Michel Monnerie's sceptical approach to witness reports and his proposal that a subjective explanation is so often found to be viable that we should apply it in all cases except those of simple misinterpretation. Like Hendry, he is able to produce any number of instances in which witnesses added unnatural details of their own to sightings whose natural basis could be positively established. But whereas to Hendry this was simply one of the many stumbling blocks the UFO investigator must learn to watch out for, the more philosophically-minded Frenchman was not content to leave it at that. He wanted to know *why* witnesses were doing this, and what was the process involved.

His first steps followed in those of Pierre Guérin, in supposing that the 'vehicle' for the spurious sighting is some kind of waking dream of the kind we considered in Chapter 3.4.

> I know of only one situation in which objects permit themselves such liberties with the laws of physics, to transform themselves from one thing to another, to multiply themselves and indulge in all manner of fantasies: the dream. By which I mean what one 'sees' whenever the subconscious takes precedence over the conscious, whether it is a simple illusion or the most complex hallucination. [105]

He emphasizes that there is nothing pathological about experiencing a waking dream of this kind. They occur, for the most part, when the individual's attention is relaxed — for example, when carrying out some banal task, taking the dog for a walk, closing the window-shutters, driving on a monotonous road, and so on; when performing an action that demands no conscious thought and so leaves the mind free to drift; or alternatively, at a critical moment of anguish when the mind craves a respite from reality.

As we saw, where Monnerie parts company from Guérin is that he sees no reason to suppose that the origin of the 'waking dream' need be sought elsewhere but in the mind of the witness. The reason why the dream takes the form it does is because the dreamer is influenced by the prevailing myths of his time. The myth of extraterrestrial entities is, in Monnerie's phrase, an 'authorized myth', carrying the sanction of social acceptance — 'perhaps the only one, in our rationalist and technological epoch, when phantoms and fairies, goblins and reiigious visions no longer arouse the same interest or attract the same belief as they once did. Human beings suffer from solitude like any intelligent species, and having once peopled space with gods, demons and spirits, they now populate it with extraterrestrials more highly developed than ourselves, onto whom we project our hopes'.

His account of the process involves an initial stimulus to act as trigger, allowing the imagined details to surge up from the subconscious:

> As we are talking about a collective myth, even those who claim not to believe in UFOs can none the less see them; moreover a group of people, influencing one another, can share a collective vision, though in such cases the details may vary from one witness to the next. In extreme cases, which do perhaps touch on the pathological, the witness, fascinated by the object which he is unable to identify, drops into a state of self-hypnosis, and can even attain a state of trance, more or less deep: it is in such cases that we have a report of contact with entities.

Without question, Monnerie's model would account for several of the anomalies of entity sighting. It would explain, for one thing, why those who go hunting for ghosts, or skywatching for UFOs, are almost always disappointed: they are in the wrong frame of mind, too alert, not sufficiently relaxed for the dream-state to occur.

It would explain, too, a particularly baffling aspect of many cases, the 'missing time' and 'empty road' phenomena, such as are displayed in a great many UFO cases, particularly those containing an alleged abduction. In the classic 'Antonia' case[125] for example, the witness, an elderly lady of the most patent sincerity, insisted that she had had a highly detailed adventure with space beings, lasting an hour or two, on a road where it was easily established that a car passed on average every half minute. If the lady had been in a dream-state, the anomaly vanishes: and the conditions — driving alone at night — were of the kind Monnerie suggests are the most conducive to being in such a state.

For these and other reasons, the waking dream hypothesis is undoubtedly very convenient. But as Belgian ufologist Jacques Scornaux pointed out when reviewing Monnerie's book, Monnerie uses the waking dream as a handy *deus ex machina*, but without being too specific as to what is meant. Evidently it involves some kind of altered state of consciousness: and after all we have observed in the course of this inquiry, we will surely find this acceptable, for a great many of our phenomena seem to have occurred when the percipient was in some kind of alternative state.[139]

But it is by no means certain that any such state as Monnerie supposes exists in fact. The closest would be a state of 'relaxed reverie', which certainly does exist insofar as it qualifies for inclusion in the *Penguin Dictionary of Psychology*, where it is defined as 'more or less aimless trains of imagery and ideas, often of the nature of day-dreaming or phantasy'. Both these terms are then in turn defined: 'day-dreaming' is described as a kind of phantasy, which is itself described as 'a form of creative imagination activity, where the images and trains of imagery are directed and controlled by the whim or pleasure of the moment.'[33]

Such a description hardly seems adequate for the state in which Monnerie alleges that someone, thinking of nothing in particular, allows his subconscious mind to persuade him that he is looking at an alien spacecraft or a group of extraterrestrial entities. There is not much 'whim' and 'pleasure' about most such reports. We must suppose that something more specific is required, something to heighten the state of mind of the witness, if we are looking for a subjective interpretation; or some external stimulus, if we are willing to contemplate such.

Pierre Guérin, as we have seen, is not simply willing to contemplate the notion of external induction; he believes it to be an essential element, concluding that a sighting does indeed represent a waking dream on the percipient's part, but that it has been artificially induced by the UFO entities, making use of a combination of personal elements drawn from the percipient's own subconscious and of the elements associated with the phenomenon itself. By putting forward a hypothesis which requires an external agent, Guérin certainly saves us from having to presuppose that the percipient is in a psychological state for which there is no precedent or parallel: but of course he also compels us to accept the reality of a far greater unknown. He requires us not only to acknowledge the existence of alien intelligences, but also to believe that they are so concerned to convince a human witness of their existence that they will go to the trouble of inducing a waking dream, rich with specific detail, which means that we have to suppose real entities who are responsible for inducing spurious images of themselves.

No doubt we must expect strangers from space to move in very mysterious ways, but the personalized induced dream hypothesis would be a difficult one to accept even if we were confident that those strangers existed. In the absence of any convincing evidence to that effect, perhaps we should take a second look at Monnerie's line of approach.

This is precisely what Monnerie himself did when, perhaps influenced by the critics of his first book, he wrote a second[106] in which he modified his hypothesis, which in its new form went something like this.

When a witness sees an object he cannot identify, he is liable to wonder whether it is perhaps one of these UFOs he keeps hearing about. So, using the details that have lodged in his subconscious mind regarding the UFO myth, he assimilates what his senses report with what he feels it must actually be. If his emotions are strongly affected, his reason switches off entirely, and he plunges into a secondary state of consciousness, something akin to delirium. But because he is 'living' the experience, it does not seem unbelievable to him, he is unaware of its inherent strangeness.

Triggering this process there must initially be some object that is genuinely not recognized by the percipient (except in rare cases where outright hallucination is involved). It is because there is, indeed, an original stimulus that is truly puzzling, that the subsequent investigator has difficulty in sorting out what is real and what was imagined. For what has taken place has been a *process of grafting* — grafting of unreal onto real, and carried out so neatly that the joins are not apparent. To the observed details will be added others, sometimes just to modify it — changing it from a star shape to a 'cigar', perhaps — but sometimes sufficient to transform it into an overwhelming experience in which the subconscious seizes its opportunity, as it were, to present a spectacular dramatization of personal hopes and fears, collective anxieties and preoccupations, using material drawn from sources some of which may lie deep in the percipient's own nature, others of which may be of external origin, subconsciously assimilated by the percipient.

Formulated in these terms, Monnerie's speculations can be seen to have considerable potential for this study, even though in the specific field for which he intended them they have failed to convince most ufologists of their viability. The process of grafting 'fiction' onto 'fact' is one way in which our entity percipients may cross the dividing-line between reality and fantasy.

3.9   THE MOTIVATED SELF

One of the themes that has worked steadily through this study as through
any detective story is the question of motive. *Why* do our percipients
have these experiences? In our final chapter we shall see what part
motivation plays in shaping the entity experience — how it causes some
to occur in the first place, and how it modifies others. First, though,
let us consider what factors are operating to cause these effects.

Motivation provides the most clear-cut distinction between the
various kinds of experience we have been considering. It is evident that
while some percipients create their entities, others have entities thrust
upon them. True, we shall not always find it a simple matter to decide
into which of these two categories some of our cases should go: but
the distinction itself is clear enough. The operative yardstick is that of
need: when need is apparent, we have no reason to look farther for
motivation.

We have already noted this difference in the case of apparitions. The
oriental entity who appeared to the lady climber in difficulties was
clearly summoned by need, the need of the percipient; whereas the
housemaid who appeared on the staircase to a houseguest meant nothing
to the percipient. If any need was operating here, it must have been
some unknown need on the part of the apparent. There are other cases
in which we *do* know the apparent's need. Here is an example.

In September 1872, Mrs Henrietta Pigott-Carleton was staying with
her husband, her father and some friends at a shooting lodge in Ireland.
One afternoon, the weather being extremely fine, she went for a country
walk with some friends, and sat reading while they occupied themselves
with fishing: her husband had some business to attend to and her father,
Lord Dorchester, stayed at the house. Unexpectedly, a storm-cloud
appeared over the mountains opposite, and she foresaw that they would
be soaked: this, she knew, would worry her father, who was always
anxious on her behalf. However, there was no shelter nearby, and
though they started back towards the house they were overtaken by
the storm and were drenched, 'looking as though we had all been barely
rescued from a watery grave'. Then, to her astonishment, she saw her
father, her husband and some other men coming towards them. 'It
seemed to me singular, not to say absurd, that my father should have

turned himself and party out in such weather. Still more to my surprise, my father evidently could not get over his disturbance, spoke little that evening, and went off to bed earlier than usual.' Next day he told her that, the previous afternoon, while sitting reading, he had seen her standing at the window, and called out to her, 'Hullo, back already?' When she didn't reply, he got up, but found no trace of her, nor had anyone else in the house seen her. He then saw the storm cloud, and became uneasy: just then his daughter's husband returned, so, despite the storm, they set out to search for her. [61]

Most such cases involve something more serious than getting soaked, but this one is interesting because a specific action was taken in consequence of seeing the apparition. The majority of phantasms of the living, when any explanation is forthcoming, tend to be apparitions of people who, at about that time, die or suffer an accident in a distant place; but only very rarely do the percipients take any action as a result. Clearly, in this case, the apparition was caused by the daughter's knowledge of the concern her father would feel; it does not appear, though, that even subconsciously she wished him to come out and 'rescue' her.

Another kind of self-evident motivation is seen in those cases where the apparition comes to warn the percipient of something that the percipient could not be expected to know. Here is a particularly striking instance:

In 1894 a Russian engineer, Vincent Zdanovich, went to buy a fur coat from a tailor named Sirota: the tailor told him that he had an almost-new one, just purchased from a merchant named Lassota, which he could have for a bargain price of 45 roubles. Pleased with the opportunity, Zdanovich bought the coat and took it home. That night he was awakened in the night and saw a stranger in his room, dressed in black: the apparition identified himself as a Mr Wischnevsky, and said 'I come to advise you to return, as quickly as possible, the fur which you have just bought, since it did not belong to Mr Lassota, but to a magistrate who has just died of phthisis. The fur is infected by phthisic bacilli.'

Though the apparition was so natural looking that Zdanovitch took it for a real person, he soon discovered his mistake, and diagnosed a hallucination. He took no action about the coat. The following night, however, he was sitting up talking with his brother Ivan, when the apparition reappeared, opening the locked door in a natural manner: this time it said, 'You are both awake. Well, this time, Mr Vincent, you will not say that my appearance yesterday was a hallucination.' He then repeated his warning, and provided some details to establish his own identity.

The following day Zdanovich returned the coat to the tailor. It was found that it had been bought from another dealer, and that the tailor had been misled as to its previous owner. The magistrate named had indeed died of phthisis, but there was some confusion over the identity of the apparent himself, perhaps due to the fact that Wischnevsky is quite a common name in that region. [148][g]

In this case the motivation underlying the incident is self-evident. What is *not* evident, though, is why it should have been this Wischnevsky — who, whoever he was, seemed to have not the slightest connection with the percipient — who nonetheless performed the role of 'guardian angel' as Zdanovich not unreasonably named him.

Apart from such specific cases, where the motive is clear-cut, there are others in which a more generalized motivation can be discerned. The difficulty which often arises here is that the state of mind embodying that motivation is likely to have persisted for a considerable period, so we have to ask what happened to make the persistent need suddenly become urgent, and manifest in the form of an entity-seeing experience? An obvious example is the sudden appearance of childhood companions in answer to what we may reasonably suppose to have been a long-felt want. We shall consider this problem in the following chapter, when we discuss the mechanism of entity sighting as a whole. What concerns us here is the nature of that persistent state.

It is evident that, in broad terms, there are social and cultural forces which will drive different people in different directions according to their individual response. One such example is millenarianism — the belief that a Golden Age is just around the corner, or would be if certain social readjustments were made. In his brilliant account of this social force and its periodic manifestations throughout history, Norman Cohn[18] shows to what lengths it can drive people — those with the appropriate disposition becoming leaders, others becoming their followers, in an infinite range of ludicrous practices in which a great many bizarre occurrences are manifested — the seeing of visions being just one of many.

The millenarianist theme is certainly part of the inspiration for many individual visionaries. When the Virgin Mary was seen at Pellevoisin in 1876, she told the percipients 'I have chosen the humble and the feeble for my glory'. It is surely not insignificant that the majority of those who claim to be accorded visions of the Virgin are humble people, usually poor children in a rural setting: a sudden glimpse of their lowly situation in the social structure might be sufficient to induce a compensatory experience in which they suddenly became the recipient of privileges, and it would be natural for this to be justified by the assertion that it was precisely for their lowliness and their humility that they were chosen — a concept which the Church welcomes.

A similar pattern obtains for those who are currently the percipients of alien visitors, who choose not to land their spacecraft on the White House lawn or on the forecourt of the Elysée Palace, but pick out ordinary housewives, cab-drivers and under-privileged persons. In the accounts the percipients give, individual problems and anxieties are not difficult to detect: for some, the aliens have come as emissaries of Satan, for others, such as George Hunt Williamson and co-author John McCoy, they are here to save us from a worldwide conspiracy in which the Jews and the communists have joined forces to destroy civilization as we know it. [167]

Because the motivation for the experience involves both personal and cultural themes, there are an infinite variety of combinations, with the result that each experience has its own individual character. The fact that the experiences are personalized in this way gives them an intensity and vividness that impresses everyone, not least the percipient himself, whose conscious mind is unaware of the forces bubbling inside him. Only when his experience is analyzed do we see how the public and private threads are interwoven, and it is more than likely that the percipient himself will refuse to accept the analysis.

The only way in which the percipient's motivations can ultimately be identified would of course be to subject him to a psychological analysis. By discovering his personal hopes and fears, his guilt feelings and doubts, his expectations and anxieties, and lining these up with his attitude towards public issues such as nuclear weapons or the presence of other ethnic groups in his community, we would be able to relate his specific sighting with his specific state. In the case of Christine Beauchamp, which we studied in the chapter on Multiple Personality, Morton Prince had the rare opportunity of making a detailed analysis of the psychology of a visionary, and he was able to conclude:

> The vision of Christ smiling was plainly a fabricated visual symbolism; it may be taken as a message sent by subconscious processes to her anxious consciousness to allay her anxiety. The phenomenon as a whole was a message addressed to her own consciousness by subconscious processes to answer her doubts and anxious questionings of herself, and to settle the conflict going on in her mind. [121]

If only we had such detailed studies of other percipients, we would doubtless be able to establish their motivation more clearly. Unfortunately, such opportunities are rare.

However, the majority of entities are not especially subtle creations, and plainly reveal in broad terms the role they are playing in the percipient's life. The demons are scary, the Cosmic Brothers are reassuring, and both tend to look the part. Visions of the Virgin show her to be comforting and kindly, but at the same time with a mother's sternness — she knows what is best for her flock, and insists that they behave themselves before they obtain her favour. Extraterrestrial aliens, too, though in general well disposed towards Earthpeople, despite our wickedness and stupidity, are determined not to help us until we start helping ourselves.

In the specific recommendations these entities make, the percipient's personal biases are often detectable — many extraterrestrials are vegetarian, for example, but some are anti-Semitic, nearly all are anti-capitalistic: some are solidly Christian while others condemn our established religions. Prejudice is revealed, too, when the entities mention specific Earthpeople as exemplars: the UMMO visitors listed a high proportion of radical heroes (Gandhi, Guevara, Martin Luther King, Marx, etc.) which, whether or not we approve their choice, does suggest a somewhat Earth-oriented political commitment; [128] a favourite of mine is American contactee Beti King, whose alien contacts have recruited not only leaders like Gandhi and Eisenhower but also American show-business personalities like Tommy Dorsey, Eddie Duchin, Louis Armstrong and Nat King Cole! [81] As mentioned earlier, when the Virgin Mary appeared in 1982 to Blandine Piegay, she not only asked that services be said once more in Latin — a much-debated theological issue of the day — but told Blandine that she must not eat so many bonbons. [47]

Given that entities are frequently expressions of the percipient's preoccupations, the question remains: what is the motivation that inspires the subconscious mind — or who/whatever — to choose this particular mode of resolving it?

In a certain number of cases, we can see that status enhancement plays a part. Religious visions are the most public and respected form of entity-sighting, and it is surely significant that they occur predominantly in cultures where to see a vision is regarded as a sign of grace, in which many would like to share even vicariously. Almost overnight, a humble peasant girl with no special attributes can become a national celebrity, as witness the crowds that throng not only to such well-established sites as Lourdes and Fatima, but to the scene of virtually any such manifestation, no matter how bizarre, and often despite the explicit disapproval of the Church itself. Blandine Piegay, again, is an example of this; her humble rustic home was invaded by thousands of pious sightseers who hoped to win some kind of spiritual comfort or reward simply by being in her presence. It is surely no accident that the great majority of religious visionaries are girls for whom society, even today, provides fewer opportunities to become 'somebody' than it does for males.

However, for the great majority of entity percipients these considerations do not apply. The rewards are private, and often amount to nothing more than the resolution of an immediate and personal problem. This can take various forms. The entity may simply serve as an exteriorization of the percipient's concern — a scapegoat for his anger against society or his sense of failure in himself, a symbol of all that he loathes or that he yearns for, a projection of his doubts or difficulties. By giving them a tangible form *outside* himself, he is confronting his problem, or dissociating himself from it, or reducing it to a form in which he feels able to handle it — the motivation will differ according to his immediate circumstances, but the process is the same.

Looked at in this light, we can see that in a great many cases entity-sighting can be interpreted as a kind of self-administered psychotherapy, and we can analyze its motivation accordingly. Some of these operations are negative, in the sense that they are no more than short-term solutions to an urgent problem, a temporary shelving of a difficult situation that will have to be tackled later when the percipient is better able to cope. Others, though, are positive, constructive, a means whereby the percipient is able to help himself, creating an entity which, admittedly, is only relaying to him a message that already exists within his own head but that his conscious mind will accept more readily if it believes that the message originates from an external authority. Thus we may reasonably suppose that Blandine Piegay was already aware that she was over-fond of bonbons; but she needed the imagined authority of the Blessed Virgin Mary to compel her to cut down on them.

Clearly, no blanket motivation for entity perception is viable: each experience is a specific response and specifically motivated. Nor must we make the mistake of assuming that all entity-sightings are strongly motivated, that this represents an extreme solution to a situation of unusual urgency. The fact that some entity-sightings occur in crisis situations, and are easily analyzed as clear expressions of a need, or as responses to urgent emergency circumstances, must not blind us to the fact that a great many others, perhaps a majority, are almost accidental and random-seeming in their occurrence — a reverie while relaxing, a briefly glimpsed apparition seen while performing some mundane task. Even in such cases, of course, it is unlikely that the nature of the experience will be wholly arbitrary; but the causes may be deep and obscure, and not necessarily related to any personal trauma at all.

So, while it may be legitimate to establish a motivation for many instances, and draw conclusions from them, we have no right to suppose that we can then proceed to project those conclusions as though they applied to *every* case. Entity-sighting may be a means to many different ends, and the resolving of personal problems may well be only one of them.

3.10 ASSIMILATED EXPERIENCE: 'THE TERROR THAT COMES IN THE NIGHT'

Throughout this study we have seen the sharp edges of supposed categories eroded, distinctions blurred; only by imposing artificial parameters have we been able to separate one kind of experience from another, treating hypnagogic experiences as if they formed a distinct class and visions of the Virgin as though they are by definition unique. However, as I trust the reader has appreciated, these distinctions were retained for the sake of convenience in discussion only, whereas it is a central point of this inquiry to consider whether our experiences may form a continuous spectrum, with specific forms resulting from specific individual or social contexts.

This is particularly true of a phenomenon we could well have considered under hypnagogic experiences, or demons, or folklore, or hauntings, or UFO-related entities: in short, a thoroughgoing cross-category experience. By a fortunate chance, too, it relates specifically to the anecdote with which I opened the book, the case of Glenda, which makes it particularly appropriate as the prelude to our concluding section. This is the type of experience frequently referred to as the 'bedroom visitor' or 'bedroom invader'; the label 'Old Hag' is however preferred by the scholar who has given it his particular attention, David Hufford, the title of whose book I have borrowed for the heading to this chapter. [75] In fact, I must acknowledge that not just the heading but the entire content of this chapter is based on Hufford's brilliant trail-blazing work, though he is not responsible for the inferences I have made or the conclusions I have drawn.

The phenomenon with which Hufford was concerned was that of a 'presence', more or less human in character, sometimes visible and sometimes invisible, which manifests to the percipient in a bedroom or similar situation and seems to oppress him physically as well as psychologically. In a questionnaire carried out among college students in Newfoundland, Hufford found that some 23 per cent had actually had some such experience, and of these, some 70 per cent knew of its significance either from personal knowledge or from hearsay. This meant that some 30 per cent had the experience *without* any such awareness: this, Hufford feels, warrants the conclusion that this type of experience has an experiential rather than a cultural source. That is to say, it represents a stereotype that can óccur to percipients irrespective of their cultural background and expectations, though their environmental circumstances might well give it a specific interpretation and even a specific label.

Subsequent research confirmed both the wide prevalence of this kind of experience and its independence of a specific cultural model. Hufford's findings also revealed the Old Hag as a 'well-kept secret'; most of those whose stories came to his attention had not publicly recounted their experiences. This Hufford reasonably attributes in large part to the fact that there is no widespread concept to which individual percipients could relate their experience, as there is for spirits of the dead, say, or alien visitors. Many of his informants prefaced their accounts by telling him that it was only when they heard him lecture or broadcast that they realized that their personal experience had been shared by others and constituted a recognized phenomenon.

The basic scenario of the experience is largely implicit in the phrase 'bedroom invader', which is why I prefer it as a label. Typically, the percipient is lying down in bed or on a couch, though not necessarily at night. Though he may have been asleep, he generally asserts strongly that it was a waking experience, even though he may also have tried to explain it to himself as some kind of dream. In those cases where observers have been present, the percipients are generally reported as having their eyes open.

In many cases, the entity is heard approaching, from beyond the room; sometimes it is seen approaching the bed, but this is not necessarily seen realistically — more than one account describes the entity as approaching from a much greater distance than separates the percipient from the door or wall. As for the entity itself, it is almost always quite anonymous, and when a face is seen it is often featureless: many times only a head is seen, sometimes a half-length body, sometimes a whole figure, sometimes a clear-cut shape, sometimes a vague blur. Sometimes the entity is felt to be more animal than human, sometimes it is no more than 'a murky presence'. It is however nearly always felt as unpleasant and menacing, and generally as quite powerful though not overwhelmingly so. It often speaks, frequently addressing the percipient by name, and perhaps using some such phrase as 'You knew that I would come.' The voices, like the entities themselves, can be of either sex, and there does not seem to be any tendency for female percipients to see male entities, or vice versa, rather than the reverse.

Hufford stresses the *conviction of reality* felt by the great majority of percipients. They often feel their bed sink beneath the entity's weight, and feel it press down on their bodies. Despite the obvious similarity to a sexual attack, this is not an aspect which is much stressed: inevitably the idea of rape is often present, but parallels with the incubus and succubus phenomena do not mean that all bedroom visitors can be seen as such. Hufford reasonably concludes that any kind of psycho-analytical explanation, such as that of Ernest Jones[77] who would certainly ascribe all such experiences to sexual repression, is simply not convincing in the light of the accounts he has collected.

Hufford is careful not to impose on the phenomena interpretations that derive from any specific belief-system; in particular he is naturally anxious to discourage any scenario with supernatural or occult implications. While I sympathize with his caution in this respect, there are certain objective features in his accounts which must be faced before the true nature of the experience is understood, and for the purposes of this inquiry we must pay them more attention than Hufford felt obliged to. These relate to the seemingly physically real components of the experience. Not only do the percipients continually stress the 'realness' of their experience, but their subjective impressions are supported by other witnesses. In one case where the percipient felt that her bed was being shaken by the entity, others in the house testified to the objective character of the events — 'it sounded like somebody was throwing our bed around': in another, a lady, babysitting for her five-year-old niece, heard footsteps approaching the room in which moments later the child would have the experience. Old Hag incidents have frequently been reported in association with hauntings and poltergeist cases, too, where there are material aspects to the business. Consequently any explanation that seeks to reduce the experience to some kind of dream or other purely mental event, must also find a way to account for these seemingly parapsychological components.

Even among the relatively small selection of cases Hufford narrates in detail, there are some that contain other features linking them with other apparitional cases, and so help to tie in bedroom invaders with the rest of our material. In one case the entity, though not an apparition of the percipient's mother, had nevertheless chosen to dress itself in one of her mother's dresses: in another, the figure of the percipient's grandmother appeared as a 'protector' at the height of the experience, and seemingly by its presence alone drove the invader away. Such accounts do not support any particular theory of apparitions, for the dress and the protecting grandmother alike could have been 'supplied' by the percipient's own mind. But they do confirm the impossibility of regarding the bedroom invaders as an independent category capable of being dealt with by the psychologist or the folklorist in isolation.

After reviewing a variety of selected cases, Hufford unequivocally asserts the wide relevance of these experiences. 'We may conclude that the features of the Old Hag attack are such that they are easily assimilated to accounts of haunting . . . the basic experience can be easily assimilated to witchcraft beliefs . . . the Old Hag can be as easily assimilated to UFO beliefs as it can to vampirism, witchcraft, or anxiety neurosis.' Consequently I feel justified in proposing that what we have, in the Old Hag/Bedroom invader experience, is a basically undifferentiated phenomenon which obtains its differentiation from individual and social circumstances.

Hufford, writing as a folklorist, is concerned to identify the fundamental underlying phenomenon, which he shows can occur in cultures ranging from the Eskimos and the Philippines to sophisticated college-educated environments. And it is precisely this masterly demonstration of a widespread phenomenon, which cannot be explained away either as a cultural artefact or as a sleep phenomenon whose content is irrelevant, that encourages me to propose that *all our varieties of entity experience can be seen as 'mechanisms', whereby hidden forces are able to manifest.* It is beside the point whether those forces spring from within the percipient, or are shaped or even initiated by external agencies. The point is that in the bedroom invader, in the demon vision, in the alien visitor, in all our categories, we have experiences whose significance is far greater than picturesque folklore on the one hand, or behaviourist mental phenomena on the other.

3.11   THE ENTITY EXPERIENCE

Our study has told us this much about the entity experience:

★ A variety of situations and circumstances can put people into states of mind in which they are liable to have the impression of seeing non-real but real-seeming entities. Some of these states are passive, others are active.

★ Though there are differences in kind between the entities thus seen, these differences can generally be traced to the percipient's social and cultural background, and do not necessarily imply fundamental differences between the phenomena themselves. It does appear, though, that certain states of mind and/or body are conducive to certain kinds of experience.

★ There is no evidence to suggest that any of these entities are 'real' in any conventional sense, though in many instances they seem capable of a sufficient degree of autonomous action to justify us in believing that they are under intelligent control, while there are sufficient examples of multiple sightings and physical effects to suggest that some of them possess a degree of substantiality.

★ There seems to be no limit to the amount of detailed information that may be embodied in the entities, and this often indicates information that the percipient could not have acquired by normal sensory means. Some cases even suggest that no living source could have supplied the information.

★ While in most cases it is reasonable to suppose that the experience originates within the mind of the percipient himself, there are some cases where the most reasonable assumption is that the sighting is initiated by the apparent, or by some agent controlling the apparent: that is, by some source external to the percipient.

There are three 'persons' involved in an entity experience:

★ the *percipient* who sees the entity;

★ the *apparent*, whom the entity appears to be;

★ the *agent*, who causes the entity to appear.

In a great many cases it would seem that the apparent and the agent are one and the same — for instance, in crisis apparitions; and in a great many more it is likely that the percipient is also the agent, creating the entity which he himself sees.

Nevertheless, it is vital to draw this distinction, because the three roles are crucial parts of the process, and if we are to understand that process we must never lose sight of the three distinct roles, even if two or even three are being played by the same actor. In particular, we must recognize that even if the percipient is both agent and percipient, this does not mean that the two are one: it is a different part of him that is functioning in each case, and those two parts of him may be hardly less separate and autonomous than if two separate persons were involved.

The experience itself also falls into three parts:

★ the *means* whereby the entity is created on the one hand, and perceived on the other;

★ the *material* from which it is constructed and the factors that dictate or modify its design;

★ the *occasion* that triggers or induces the manifestation.

Each of these elements has been touched on intermittently in the course of our inquiry, some recurring more or less continuously; so a quick résumé should be sufficient at this stage.

The means

Our study of dissociated personality showed that there exists, within the human mind, a capability for one part of that mind to function autonomously: one of the things it can do is to stage a display, which it presents for the conscious mind to perceive. The conscious mind may have little or no idea that this is happening, and may be completely deceived by the fiction that is being presented to it.

Since we know that this capability exists, we may reasonably suppose it to be operating in cases where we can see only the end-result with no indication as to how that end-result was achieved. We are encouraged to do this by the evidence of such experiences as dreams, where it seems self-evident that the source of the experience is within our own minds, and that it is a subconscious process of which our conscious minds become aware only when the subconscious mind chooses, though perhaps sometimes occasional accidental glimpses may occur, as in momentary waking dreams.

It is reasonable to hypothesize a 'producer' for these imaginary experiences. No scientific label is available for a function unknown to physiology: nevertheless, 'something' constructs our dreams, and it is convenient to give that something a label, even if we commit ourselves to saying nothing more about it. And yet we must acknowledge the astonishing capability of the producer, which goes far beyond any mere mechanical manipulations of data. His handiwork — dreams, visions, hallucinations, etc. — displays outstanding artistry and ingenuity, far beyond our waking ability; he displays great flexibility, as exemplified in the way he will at a moment's notice incorporate into a dream the sound of a clock or some other sensory input; he often manifests purpose, presenting us with experiences that resolve our problems or ease our minds; and in doing so he reveals a deep understanding of our needs, hopes, fears and concerns. At the same time he is often extremely witty, and gives every appearance of enjoying himself in the way, for example, he devises our dreams. All this adds up to something very like a personality; and indeed, fortified by our study of dissociated personalities, we need not be timid about supposing that nothing less is in fact involved, and that our producer is indeed something akin to an autonomous personality, separate from and in many respects superior to our conscious self.

It may not always be the case, however, that the entity experience originates within our own minds. Often, for example, it contains material that could not be known to us by any normal means. In some such cases — for example, when we dream of something that actually occurs the following day — we may still credit our producer, but now we have to make the additional assumption that he has access to sources of information that are not available to our conscious minds.

However, there are certain categories of sighting — for example, crisis apparitions — where even this will hardly do, and it seems in every way more reasonable to presume that an external agent has had a part in the business. In these cases we may suppose either that the external agent takes over the producer's role altogether, or that the producer acts as a 'relay station' for the signal coming in from outside. But there is good reason to suppose that the eventual programme is the result of co-operation between our producer and the outside agent, embodying information supplied by both, so that the apparition of dying Aunt Agnes is able to appear, quite naturalistically, in nephew George's room, which Agnes herself has never visited.

The physiological mechanism whereby the experience is presented to the conscious mind is a matter we have little information about. However, a few tests have been carried out, notably on the subject Ruth, whom we considered in Chapter 2.3. These showed that when Ruth was hallucinating, her retina and pupils responded normally to the 'real' scene, but the visual evoked response in her brain *was* affected: that is to say, somewhere along the route between her eyes and her brain, the input from her senses was intercepted, and the alternative signal — in this case a deliberately induced hallucination of her father — was somehow substituted.

Evidently, then, our producer is able to override or by-pass our senses and ensure that it is his programme which is performed in the theatre of the brain, and not that supplied by the senses monitoring the real world about us. I find the analogy of the video-recorder helpful. We may think of our minds as permanently tuned to the official broadcasting station, whose transmissions come into our brain continuously via the senses; so day and night we are more or less aware of the real-time material being broadcast. But at the same time, in his private workshop, our producer is busy making video programmes of his own. Indeed, there is evidence to suggest that he never stops doing so by day or night, which is why there is alternative viewing material available at all hours, even if we lapse momentarily into a state of reverie to find curious images flitting through our minds like fragments of what is being shown on other channels glimpsed as we press the control buttons of our television set. (Possibly, though, our hard-working producer *does* take an occasional rest; perhaps that is what *real* sleep is for, the phases when we do not even so much as dream.)

Every so often, however — we shall be considering the occasions for it in a moment — a more positive act of switching occurs, and then the percipient has a sudden, lifelike experience of intense vividness. If the producer has done his job well, the fabricated programme will be so carefully adjusted to reality that the conscious mind will not be aware of the switch, and will mistake the illusion for reality. At other times, the producer may not bother, or may even choose to defy reality altogether, so that the percipient knows at once, or very soon realizes, that what he is seeing is something other than the regular transmission, something different from everyday reality.

The material

The analogy of the video-recorder continues to hold good when we turn our attention to the physical form in which the material forming the entity-experience is communicated. Just as a video-tape gives us a replay of Ibsen's *Ghosts* in which the performances of the original live actors are, as it were, enacted by encoded signals, so we know that our producer must produce *his* work in the form of encoded signals, which are fed into the brain just like the normal sensory input. We must not forget that our minds work with encoded information like any purely mechanical computer. Our memories of a summer holiday may evoke vividly nostalgic recollections of sun, sand and sea, but for all that they are brought from the memory-store to our conscious attention along channels which differ from those of the video-recorder that enables us to replay the camera-record we made of that same holiday only in their incalculably greater complexity.

Such an analogy is plausible so long as we are concerned only with internally projected material — when it is simply a case of our producer putting on a display for the benefit of our conscious minds. However, some other process must be involved when external manifestation takes place. This occurs when someone sets out to project an image of himself to a distant location where it is seen by others; and when more than one percipient sees the same entity; and possibly in those cases where the entity, though seen only by one solitary percipient, appears to grow from, say, a ball of light — a process that would hardly be necessary if all that was happening was a purely internal hallucination of the kind we have been considering.

In such cases as these, we have no choice but to recognize that some kind of material reality is involved, and that in some way it is controlled by the agent. Somehow, the agent is able subconsciously to produce and project an image — generally of his own self — and despatch it to a distant place, which may well be a place with which he is unfamiliar. Moreover, that image is endowed with senses in some way: whether it is autonomous — perhaps even to the extent of being temporarily the seat of the percipient's 'awareness' — or whether it is simply a kind of remotely-controlled robot, the projected image is undoubtedly capable of noting its surroundings, responding to them and bringing back information about them.

When we reach this point, the video-recorder at its current stage of development is no longer adequate as an analogy, but there is another technical development, still in its infancy, which is more helpful: the holograph. Here we have a device that enables a non-physical image to be projected at a distance to give a three-dimensional illusion of reality. I do not want to push this analogy too hard, because it is quite clearly appropriate only to a very limited extent: nevertheless, the very existence of the holograph is an encouragement to open our minds to the possibility that such image-projection could be effected by our minds *as a physical reality*, rather than as some occult procedure that requires us to believe in an order of activity with no basis in the physical universe as we know it.

I think it may also be worth citing two bizarre reports that seem to indicate that entities could have some kind of electromagnetic basis. That sentence is deliberately phrased loosely, because the data are far too undependable for us even to attempt an explanation of what may be going on — if, indeed, anything is going on at all. Had I known of only one such case, I doubt if I would have included it; but I have two, and here they are:

Several years ago [that is, previous to 1951] in the laboratory of the Rhodes Electrical Company, London, chief engineer Eastman was working on some high-tension wires forming a magnetic field in a dark room. Suddenly he observed a luminous blue sphere form above a nearby revolving dynamo. As the light became more intense, a form resembling a human hand appeared in the centre of the sphere. Eastman and his assistant, Harold Woodew, watched the phenomenon for several minutes before it faded away. The two men spent four days trying to recreate the conditions, and eventually obtained a human head, with indistinct features, white in colour, and revolving slowly. According to Vincent Gaddis, Eastman photographed it. [42a]

Fay Clark told Brad Steiger that one evening in 1931, at the Northern State Power Company in LaCrosse, Wisconsin, a standby boiler had just been put on line at maximum power, when a cloud began to form over the turbine. Fearing overheating, the men checked and found that the machine was operating normally. They then saw appear in the cloud 'as clearly as could be, the image of a woman lying on a couch. One of her arms was covered with jewels, and there were rings on her fingers. All of the men witnessed this for about twenty seconds before it faded out. The engineer told his companions that he had witnessed similar phenomena before in England. He said that he believed that the tempo of the electrical generator had thrown the area out of frequency with our era of civilization, that we had somehow attuned to the Past'. [152]

Given that people with a particular trade move from one job to another in their line of business, it is still somewhat of a coincidence that the same engineer should have been involved in more than one such incident, which is clearly very rare. Whether the incident in England referred to by the engineer in the second story is the one described in the first story is something we do not know, but it is possible, in which case we have to consider the possibility that it was something to do with the engineer himself — that he was the catalyst, as it were, which enabled this rare manifestation to occur.

It is interesting, too, that the engineer was so ready with an explanation that I do not think is the one which would have come first to most people's minds. However, it parallels very closely the response of Miss Moberly and Miss Jourdain to their strange experience at Versailles, when they too came to the conclusion that they had witnessed a scene from the past.

I present these unsupported anecdotes for what they are worth, which is next to nothing as scientific evidence, but perhaps somewhat more as a clue to the fundamental physical nature of our entities. It is tempting to proceed to further conjecture, but I intend to resist that temptation, and merely comment that here we have a possible loophole in the seemingly unbreachable wall that separates 'reality' from 'illusion'.

Hypothetical though these notions are, they are consistent with the facts. And since these projected images, whatever kind of physical reality they possess, must be made out of 'something', just as they must be made by 'someone', we may call that material the *psi-substance*, without attempting a more precise description. Whether it is in any way related to ectoplasm is anyone's guess, just as is the question whether it is related to the auras which some 'psychically endowed' persons claim to see surrounding others. Nor will I embark here on speculation about how any of this ties in with the question of whether we survive death, though of course it is certainly relevant.

This much, however, I believe we are justified in saying: *The entity-experience has a material basis that can reasonably be conceived of as a physical communication, fabricated by an autonomously operating part of the percipient's mind, either on its own or in liaison with an*

external agent; expressed in the same encoded-signal form as any other
mental communication; presented to the conscious mind as a substitute
for the sensory input from the real world; and occasionally being given
a temporary external expression utilizing some kind of quasi-material
psi-substance.

The content of the entities, as opposed to their form physical or otherwise, takes us away from physiology to psychology and sociology. When it is an entity whose identity is known, such as Aunt Agnes, no problem arises. It is like a documentary programme on television: our producer has to do the best he can with the material provided, and though he may do clever things with camera angles and lighting, basically what nephew George sees is Aunt Agnes as she is.

In non-veridical cases the producer gets more of a chance to display his talents. If his assignment is to produce a vision of the Virgin Mary, he will look to see what stereotype is stored in the percipient's mind, but his final design may differ from the stereotype in significant respects. Thus the vision seen by Bernadette, though ultimately identified as the Virgin Mary, differed so far from the stereotype as to make Bernadette herself question its identity. We may suppose that in this case Bernadette's producer exercised his creative abilities to endow the image with individually selected attributes. No doubt Bernadette wanted to see the Virgin Mary, though perhaps not as explicitly as Catherine Labouré did (see Chapter 1.8); but the Virgin Mary of her conscious expectation was not, perhaps, the Virgin of her subconscious yearning. So what her producer gave her was a custom-made Virgin designed to her individual specifications.

Many years ago, the American magazine *Life* presented a pair of contrasting full-page pictures of girls. On the left was 'the Editors' Choice' — an archetypal showgirl in fishnet tights, projecting bosom, blonde hair, bare shoulders and all: on the right was the one the readers had chosen — fully-dressed, lightly made-up, with an open, innocent face — everyman's image of the Girl Next Door. Something like this, I suggest, is going on in the entity-fabricating business: our conscious editorial minds think they know what we want, but our subconscious producer knows better, and creates something that reflects our unarticulated wishes, fears, hopes or preoccupations.

Nevertheless, stereotypes abound. Comforting entities tend to be 'Comrades in White' while threatening ones are more likely to be 'Men in Black'. Friendly aliens have long golden hair to their shoulders, while the unfriendly sort have scaley skins and slanting eyes. There is a generally accepted code in these matters. Whether that code can be extended to embrace Jung's archetypes from the collective unconscious, however, is not borne out by our study. It is a great and stimulating concept, and I am reluctant to discard it: perhaps those alien visitors who arrive in egg-shaped spacecraft are expressing the notion of wholeness which the egg shape is supposed to symbolize. I can only observe that you need to work hard to find archetypal components in the average entity report, and even if you find them, they are liable to be intellectual abstractions, which seem unlikely to have inspired a vivid emotional experience.

Far more important, it seems to me, are the cultural and social concepts that the percipient has acquired, not as a subscriber to the collective unconscious, but as an individual human earthling of his day and age. It is as such that he absorbs the images and concepts that resonate with his personal concerns and preoccupations — those that express the guilt-soaked remorse or the proffered solace of religion, the escape from everyday tedium offered by science fiction or space exploration, the menace of sinister foreigners or totalitarian conspiracies, the threat of disasters natural or man-made, the hope promised by millenarianist messiahs in this world or cosmic brothers from another — any of which can be metamorphosed by the producer into a meaningful element in the fabricated experience.

The occasion

What causes the entity experience to occur? Who chooses the appropriate moment, and for what reasons?

In some cases it is self-evident. When a crisis occurs in the life of the agent or the percipient — when Aunt Agnes has her accident, when the lady climber finds herself stranded on the mountain-side — clearly a suitable moment is at hand. The crisis does not have to be very severe, as we saw in the case of the lady caught in the rainstorm. Nor does there have to be an immediate, urgent danger: prolonged strain, it seems, may have the same effect, as with mountain-climbers and others whose physical hardships are relieved by the presence of imagined companions.

There are other continuous situations, such as that of the lonely child who creates an imaginary companion for herself. What is the immediate occasion that causes the entity to appear? Perhaps investigation would reveal that a harsh word from a teacher or a guardian was the last straw: but if so it was probably *only* the last straw. That is, the eventual manifestation was the consequence of a long period of gestation, akin to that of the conversion process noted in religious-minded adolescents,[76,151] in whom a build-up, which may have been extended over months or even years, results in a sudden mental explosion whose immediate cause can be of the slightest — simply entering a church on impulse may be sufficient — whereupon the whole process goes into action like a clock striking the hour.

In this connection we must remind ourselves not to jump to the conclusion that the existence of a need in the percipient's mind is the direct cause of his sighting. It is more likely that the need creates a state of mind, in which an entity-sighting becomes possible or appropriate. Under these circumstances, the trigger-incident can be of the slightest: we can, if we like, think of the subconscious mind waiting its chance and taking advantage of the opportunity, however slight, when it presents itself.

Even if we suppose an external source for some of our sighting experiences, the same conditions may apply. The induced-dream hypothesis, of which we noted several variants, requires that the percipient shall be in an appropriate state to receive it. Hence, no doubt, the preponderance of sightings that occur when the percipient is in a relaxed, passive or dissociated state. Morton Prince, Christine Beauchamp's doctor, believed that 'a certain amount of dissociation is probably always necessary', though I think this is debatable. I think perhaps Gordon Rattray Taylor was right to extent the range of suitable conditions:

> I am inclined to think that the production of imagery is going on all the time in the older, more primitive part of the brain, but is normally ignored, or shut off by the cortex. Only when the cortex is numbed or unoccupied — or when the fantasy is strongly amplified by emotional loading — does it break through into consciousness. [157]

I am not sure that he is right to attribute these experiences to a 'more primitive' part of the brain, but I think his suggestion of image-enhancement due to emotional factors is a helpful one.

Given that we need to hypothesize an external agency for any but the special categories already noted, then I think that Maxwell Cade's suggestion, formulated in respect of UFO entities but capable of wider application (see page 164), is a potentially useful one. He suggested that alien entities might have the power to scan our minds, extract the appropriate information as to our expectations, fears or hopes, and instantaneously fabricate entities that would reflect those expectations to our conscious minds, thus minimizing the shock of such an encounter.

But do we need to hypothesize an external agency at all? Strictly speaking, probably not. Given a sufficiently skilled producer, given the necessary mental means, given a sufficient motivation, given an infinite information source, given the multifarious idiosyncracies of which the human mind is capable, I think it is possible to say that we need not hypothesize any external agency except for those where it is clearly indicated, such as crisis apparitions.

But just because it *need not* be, it does not follow that it *is not*. The law of parsimony requires that, other things being equal, the simplest explanation should be adopted. But other things are never equal. If you believe that the Devil exists, then the simplest explanation for many kinds of entity is that they are the Devil or his creations. If you believe that the Virgin Mary watches over us, then when you see an entity which conforms to your notions of the Virgin, it is simplest to believe that that is what it really is. One man's 'simplest explanation' is not another's.

★   ★   ★

So are we now in a position to say what is happening when George sees an apparition of his Aunt Agnes in his bedroom; when a climber in difficulties is rescued by a mysterious oriental stranger; or when a French peasant girl receives a visit from the Madonna? In broad terms, yes, I think we are; it is something like this.

Within our minds there exists a creative, intelligent, sympathetic and understanding capability, whose function is to fabricate non-real scenes and scenarios, for purposes only some of which can be guessed at. This capability, which for the sake of convenience we may call the *producer*, may plausibly be conceived as a parallel personality to our conscious personality.

The producer has access not only to all our sensory input, both conscious and unconscious, but also to our mental and emotional attitudes and concerns; he also has access, whether constant or when the need arises, to information not available to our conscious minds, because it relates to events occurring out of our reach in space or time, including events that have not yet occurred.

Using this material, the producer creates fantasies consisting largely of representations of people; these may be persons, living or dead, known to the percipient; or stereotypes whose identity is evident though they are not personally known to the percipient, such as the Virgin Mary or an extraterrestrial alien; or persons who, so far as the conscious mind can tell, are total strangers. These may appear as isolated figures or in realistic settings: in either case the manifestation is managed so skilfully that the entity is frequently assumed to be real at the time.

There is evidence to show that the creation of these imaginary scenes and entities is a continuous process, but unquestionably some are created at specific times for specific purposes.

The non-real scenes are substituted for the reality reported by the senses; the substitution takes place somewhere between the sense organs and that part of the brain concerned with visual imagery. It is effected so neatly that there is generally no discontinuity between reality and fiction. There is no reason to suppose that it uses anything but the normal channels of mental communication, employing encoded signals.

While the most memorable instances of this process are the made-to-order experiences that relate to a crisis or other event with a strong emotional loading, 'accidental' tuning in to the material can be obtained in a number of mental states, which are not those of everyday consciousness — when intoxicated or drugged, in trance, delirium or mystical ecstasy. It seems that these states provide an equivalent of the highly charged emotional states that are generally the occasion for the experience; all, alike, *enable* it to take place.

There are also, however, cases that suggest that an external agency is responsible for the event and provides the requisite emotional drive; in such cases, all that is required of the percipient is that he be in a suitable receptive state.

There is no evident limit to the range of material of the experience, but its nature will be determined by the percipient's personal preoccupations, his cultural background, and by the immediate situation. It will also be adapted to the context of time and place in which it occurs.

A made-to-measure entity-sighting seems to be just one of several options whereby the subconscious personality can influence our conscious minds. We may surmise that it is when a visual display seems, for whatever reason, to be the most appropriate or the most effective way of presenting information to the conscious mind, that this process is selected, and our conscious selves are presented with the illusion of an entity-experience: nephew George sees an apparition of his Aunt Agnes; Bernadette Soubirous sees 'a girl in white, no taller than myself'; an endangered climber is rescued by a kindly oriental appearing from nowhere; Martin Luther is confronted by the Devil; little Eileen Garrett gets a friend to play with; and Mrs Laura Mundo is visited by Orthon from the planet Venus.

BIBLIOGRAPHY

Unless otherwise indicated, publication is in London. Wherever possible, first publication is indicated; in some cases, however, revised editions have been used.

1 Abercrombie, John, *Inquiries concerning the intellectual powers* (Murray, 1830).

[Adamski, George, *see* Leslie, Desmond.]

2 Allison, Ralph, *Minds in many pieces* (Rawson, Wade, New York, 1980).

3 Bagnall, Oscar, *Origins and properties of the human aura* (Kegan Paul, 1937).

4 Barker, Gray, *They knew too much about flying saucers* (Werner Laurie, 1958: first published 1956).

5 Basterfield, Keith, *An in-depth review of Australasian UFO-related entity reports* (ACUFOS, Australia, 1980).

6 Bayless, Raymond, *Experiences of a psychical researcher* (University Books, New York, 1972).

7 Bender, Albert, *Flying saucers and the three men* (Saucerian Books, Clarksburg, West Virginia, 1962).

8 Bennett, Sir Ernest, *Apparitions and haunted houses* (Faber, 1939).

9 Bisson, Juliette Alexandre, *Les phénomènes dits de matérialisation* (Felix Alcan, Paris, 1914).

10 Blum, Ralph and Judy, *Beyond Earth* (Bantam Books, New York, 1974).

11 Branden, Victoria, *Understanding ghosts* (Gollancz, 1980).

12 Brennan, J. H., *Experimental magic* (Aquarian Press, 1972).

13 Broad, C. D., *Lectures on psychical research* (Routledge, 1962).

14 BUFORA (British UFO Research Association) *Journal*, Vol. 8, No. 4, (1979).

15 Cade, Maxwell, 'A long cool look at alien intelligence', in *Flying Saucer Review*, Vol. 13, No. 6, (1967).

16 Cavendish, Richard, *The black arts* (Routledge and Kegan Paul, 1967).

17 Chibbett, H. S. W., 'Ufos and parapsychology', in *Flying Saucer Review Special: UFO percipients* (September 1969).

18 Cohn, Norman, *The pursuit of the millenium* (Secker and Warburg, 1957).

19 Condon, Edward U. (director), *Final report of the scientific study of Unidentified Flying Objects* (University of Colorado, 1968).

20 Conway, David, *Magic, an occult primer* (Cape, 1972).

21 Cornu, Gilbert, *Pour une politique de la porte ouverte,* in *Lumières dans la nuit* (Le Chambon, 1981-1982).

22 Cove, Gordon, *Who pilots the flying saucers?* (Private, Lytham, *circa* 1955).

23 Crawford, W. J., *Psychic structures* (Watkins, 1921).

24 Cristiani, Léon, *Présence de Satan dans le monde moderne* (France Empire, Paris, 1959). [English version: *Satan in the modern world* (Barrie and Rockcliff, 1961)].

25 Crookall, Robert, *The supreme adventure* (James Clarke, 1961).

26 Crowe, Catherine, *The night side of nature* (Newby, 1848).

27 Crowley, Aleister, *The book of the Goetia of Solomon the King* (Society for the Propagation of Religious Truth, Boleskine, Inverness, 1904).

28 David-Neel, Alexandra, *Magic and mystery in Tibet* (University Books, New York, 1965: first published 1932).

29 De Boismont, A. B., *On hallucinations* (Renshaw, 1859: Translated from the French by R. T. Hulme).

30 DeHerrera, John, *The Etherean invasion* (Hwong, Los Angeles, 1978).

31 DeHerrera, John, *The imaginary encounter study* (The author, 1980).

32 Doyle, Sir Arthur Conan, *The coming of the fairies* (Hodder and Stoughton, 1922).

33 Drever, James, *A dictionary of psychology* (Penguin, 1952; revised 1964).

34 Eisenbud, Jules, *The world of Ted Serios* (Cape, 1968: first published in USA, 1966).

35 Englebert, Omer, *Catherine Labouré and the modern apparitions of Our Lady* (P. J. Kenedy, New York, 1959).

36 Estrade, Jean Baptiste, *Les apparitions de Lourdes* (Lourdes, Imprimerie de la Grotte, 1953: written in 1888 and first published in 1890).

37 Evans, Hilary, *'The Welsh Triangle'* in *The unexplained,* issue 41 (Orbis, 1981).

38 Evans, Hilary, *Intrusions* (Routledge and Kegan Paul, 1982).

39 Evans, Hilary, *'A night to remember',* in *The unexplained,* issue 136 (Orbis, 1983).

40 Evans, Hilary, *The evidence for UFOs* (Aquarian Press, Wellingborough, 1983).

41 Faill, Bill, in Rogo, Scott, *UFO abductions* (Signet, New York, 1980).

42 *Fate* magazine, Highland Park, Illinois.
    a — April 1951: Gaddis, Vincent, *Electrical ghosts.*
    b — May/June 1951: Haywood, Captain C. W., *The white cavalry.*
    c — Oct 1957: *Return in the night* [Travers case]
    d — June 1960: Mamontoff, Nicholas, *Can thoughts have forms?*
    e — Sept 1961: Nelson, I. M., *Menacing visions.*
    f — April 1966: Podolsky, Edward, *Have you seen your double?*
    g — March 1967: King, Peter, *Who was the fourth man?*
    h — May 1968: letter by Revd. Albert H. Baller, *The mysterious defender.*
    i — May 1969: Evans, Mayme, *I never flew alone.*
    j — Aug 1971: *Ghosts on the mountain,* in Fuller, Curtis, *I see by the papers.*

k — June 1977: the Harley Cross incident, in Fuller, Curtis, *I see by the papers.*

l — May 1982: Rogo, D. Scott, *Natuzza Evolo works miracles.*

43 Flammarion, Camille, *Death and its mystery: volume two, At the moment of death* (Fisher Unwin, 1922).

44 *Flying Saucer Review,* West Malling, Kent.
   a — Vol. 11, Nos. 5, 6 — Valensole case.
   b — Vol. 20, Nos. 2, 3, Vol. 21, No. 3/4 — New Berlin case.
   c — Vol 21, Nos. 1, 2 — Zimbabwe case.
   d — Vol 23, No. 1 — West Wales case.
   e — Vol 23, No. 4 — Hopkins case.

45 Fortune, Dion, *Psychic self-defence* (Rider, 1930).

46 Fowler, Raymond, *The Andreasson affair* (Prentice Hall, New Jersey, 1979), and *The Andreasson affair, phase two* (1982).

47 *France Soir,* 13 avril 1982.

48 Fuller, John G., *The interrupted journey* (Dial Press, New York, 1966).

49 Gardner, Edward L., *The theosophic view of fairies,* in Doyle (note 32).

50 Garrett, Eileen, *My life,* (Rider, 1939).

51 Geley, Gustave, *Clairvoyance and materialisation* (Fisher Unwin, 1927).

52 Gibbons, A. O., (ed), *The Trianon adventure* (Museum Press, 1958).

53 Goethe, Wolfgang von, *Aus meinem Leben,* cited in Shirley, Ralph, *The mystery of the human double,* (Rider, *circa* 1930).

54 Goldenson, Robert M., *Mysteries of the mind* (Doubleday, New York, 1973).

55 Görres, *La mystique divine, naturelle et diabolique,* translated into French by Saint-Foi (Paris, 1854) [Perez case is in Vol. 4].

56 Goss, Michael, *On the road again,* in *Fortean Times,* issue 34 (1981).

57 Goss, Michael, *The evidence for the phantom hitch-hiker* (Aquarian Press, Wellingborough, 1984).

58 Green, Celia, and McCreery, Charles, *Apparitions* (Hamish Hamilton, 1975).

59 Guérin, Pierre, 'Le problème de la preuve en ufologie', in Bourret, J. C., *Le nouveau défi des OVNI* (France Empire, Paris, 1976).

60 *Guide de la France mystérieuse* (Tchou, Paris, 1964).

61 Gurney, Edmund; Myers, F. W. H.; and Podmore, Frank, *Phantasms of the living* (Society for Psychical Research, 1886).
   Case No. 7 — Beard
         No. 12 — Wesermann
         No. 13 — Moses
         No. 14 — Beard
         No. 25 — Collyer
         No. 146 — Brougham
         No. 194 — Miss R. and the Captain
         No. 215 — Rouse
         No. 242 — Clerke
         No. 321 — Gwynne
         No. 578 — Piggott-Carleton
         No. 686 — Russell

Vol. 2, p. 206 (unnumbered) — Becket

62 Haisell, David, *The missing seven hours* (Paperjacks, Markham, Ontario, 1978).

63 Halifax, Lord, *Ghost Book, volume one* (Bles, 1936).

64 Hall, Trevor H., *The spiritualists* (Duckworth, 1962).

65 Harner, Michael J., *Hallucinogens and shamanism* (Oxford University Press, New York, 1973).

66 Hart, Hornell, 'Six theories about apparitions', in *Proceedings of the Society for Psychical Research*, Vol. 50 (May 1956).

67 Head, Henry, *Certain mental changes that accompany visceral disease*, Goulstonian Lecture, reprinted from *Brain*, Vol. 24, Part 111 (Bale Sons and Danielsson, 1901).

68 Hendry, Allan, *The UFO handbook* (Doubleday, New York, 1979).

69 Hibbert, Samuel, *Sketches of the philosophy of apparitions*, (second edition, Oliver and Boyd, 1825).

70 Hilgard, Ernest R., *The divided consciousness* (John Wiley and Sons, New York, 1977).

71 Holms, A. Campbell, *The facts of psychic science and philosophy* (Kegan Paul, 1925).

72 Home, Daniel Dunglass, *Lights and shadows of spiritualism* (Virtue, 1877).

73 Hopkins, Budd, *Missing time* (Richard Marek, New York, 1981).

74 Howitt-Watts, Anna Mary, *Pioneers of the spiritual reformation; life and works of Dr Justinus Kerner* (Psychological Press Association, 1883).

75 Hufford, David J., *The terror that comes in the night* (University of Pennsylvania Press, Philadelphia, 1982).

76 James, William, *The varieties of religious experience* (Longmans Green, 1902).

77 Jones, Ernest, *On the nightmare* (Hogarth Press, 1931).

78 Jung, Carl Gustav, *The archetypes and the collective unconscious* (Routledge, 1959 [first published 1954]).

79 Jung, Carl Gustav, *Flying saucers* (Routledge, 1959).

80 Kilner, Walter J., *The human atmosphere* (Kegan Paul, 1920: an earlier version appeared in 1911).

81 King, Beti, *Diary from outer space* (private, USA, 1976).

82 Klarer, Elizabeth, *Jenseits der Lichtmauer* (Ventla, Wiesbaden, 1977). [English version: *Beyond the light Barrier*, Howard Timmins, Cape Town, 1980].

83 Lagarde, Fernand, in *Lumières dans la nuit*, issue 81 (1967).

84 Lagarde, Fernand, *Commentaires sur la Dame Blanche*, in *Lumières dans la nuit*, issues 213/214 (1982).

85 Langton, Edward, *The supernatural* (Rider, 1934).

86 Langton, Edward, *Satan, a portrait* (Skeffington, no date, *circa* 1945).

87 Lavater, Lewes, *Of ghosts and spirits walking by night* (Shakespeare Association and Oxford University Press, 1929: first published 1572).

88 Lawson, Alvin, *What can we learn from hypnosis of imaginary UFO abductees?*, at MUFON Symposium, Arizona, 1977.

89 Lawson, Alvin, 'The hypnosis of imaginary UFO abductees', in Curtis Fuller

(ed.), *Proceedings of the first international UFO congress, 1977* (Warner Books, New York, 1980).

90 Lawson, Alvin, *Alien roots: six UFO entity types and some possible earthly ancestors*, at MUFON Symposium, California, 1979.

91 Leaning, F. E., 'An introductory study of hypnagogic phenomena', in *Proceedings of the Society for Psychical Research*, Vol. 94 (May 1925).

92 Le Bec, *Medical proof of the miraculous* (Harding and More, 1922).

93 Leslie, Desmond, and Adamski, George, *Flying saucers have landed* (Werner Laurie, 1953).

94 Lévi, Eliphas, *Dogme et rituel de la haute magie* (Baillière, Paris, 1861).

95 Lorenzen, Jim and Coral, *UFOs over the Americas* (Signet, New York, 1968).

96 'Quasi-atterrissage près de Missancourt', in *Lumières dans la nuit*, issue 149 (1975).

97 McClure, Kevin, *The evidence for visions of the Virgin Mary* (Aquarian Press, Wellingborough, 1983).

98 Mace, C. A., 'Supernormal faculty and the structure of the mind', in *Proceedings of the Society for Psychical Research*, Vol. 44, (1937).

99 Machen, Arthur, *The bowmen* (Simpkin Marshall, 1915).

100 Mackenzie, Andrew, *Hauntings and apparitions* (Heinemann, 1982).

101 Martindale, C. C., *The message of Fatima* (Burns, Oates and Washbourne, 1950).

102 Masters, E. L., and Houston, Jean, *The varieties of psychedelic experience* (Blond, 1966).

103 Méheust, Bertrand, *Science fiction et soucoupes volantes* (Mercure de France, Paris, 1978).

104 Moberly, S. A., and Jourdain, Frances, *An adventure* (Macmillan, 1911: first published anonymously).

105 Monnerie, Michel, *Et si les OVNIs n'existaient pas?* (Les Humanoïdes Associés, Paris, 1977).

106 Monnerie, Michel, *Le naufrage des extraterrestres* (Nouvelles Editions Rationalistes, Paris, 1979).

107 Moss, Peter, *Ghosts over Britain* (1977).

108 Myers, F. W. H. *Human personality* (Longman, 1903).

109 Naranjo, Claudio, 'Psychological aspects of the Yagé experience in an experimental setting', in Michael Harner (ed.), *Hallucinogens and shamanism* (Oxford University Press, New York, 1973).

110 Neher, Andrew, *The psychology of transcendence* (Prentice Hall, New Jersey, 1980).

111 Nider, Johannes, *Formicarius* (1692), cited in Harmer (ed.), *Hallucinogens and shamanism* (Oxford University Press, New York, 1973).

112 Oesterreich, Traugott K., *Possession*, translated from the German by Ibberson (1921), reprinted [as *Possession and Exorcism*] by Causeway Books, New York.

113 Olliver, C. W. A., *The extension of consciousness* (Rider, 1932).

114 Owen, Iris M., with Sparrow, Margaret, *Conjuring up Philip* (Fitzhenry and Whiteside, Ontario, 1976).

115 Owen, Robert Dale, *Footsteps on the boundary of another world* (Lippincott, Philadelphia, 1860).

116 Parish, Edmund, *Hallucinations and illusions* (Scott, 1897).

117 Petitpierre, Dom Robert, *Exorcising devils* (Hale, 1976).

118 Piddington, J. G., on Dr Head's Goulstonian Lectures (see note 60) in *Proceedings of the Society for Psychical Research* Vol. 51 (1905).

119 Podmore, Frank, *Modern spiritualism* (Methuen, 1902).

120 Price, Henry Habberley, 'Haunting and the psychic ether hypothesis', in *Proceedings of the Society for Psychical Research,* Vol. 44 (1939).

121 Prince, Morton, *Dissociation of a personality* (Longman, 1905).

122 Prince, Morton, *The unconscious* (Macmillan, 1914).

123 Prince, Walter Franklin, *The enchanted boundary* (Boston Society for Psychical Research, 1930).

124 Pugh, Randall Jones, and Holiday, F. W., *The Dyfed enigma* (Faber, 1979).

125 *Revue des Soucoupes Volantes,* No. 5 (France, 1978).

126 Rhine, Louisa, *Hidden channels of the mind* (Gollancz, 1962).

127 Rhine, Louisa, *The invisible picture* (McFarland, Jefferson, North Carolina, 1981).

128 Ribera, Antonio, *El misterio de Ummo* (Plaza y Janes, Barcelona, 1979).

129 Ribet, M. J. *La mystique divine* (Poussielgue, Paris, 1895).

130 Richet, Charles, *Thirty years of psychical research* (Collins, 1923).

131 Rifat, Claude, 'Is the *locus coeruleus* involved in the most bizarre aspects of UFO reports?' in *UFO Phenomena,* Vol. 11, No. 1 (Editecs, Milano, 1978).

132 Rifat, Claude, interviewed in *OVNI-présence* No. 25 (Génève, March 1983).

133 Roberts, Anthony, and Gilbertson, Geoff, *The dark gods* (Rider-Hutchinson, 1980).

134 Robinson, John J., 'MIB, *Cadillacs Etc.'* in T. G. Beckley (ed.), *Aliens among us* (Global Communications, New York, 1971).

135 Rogo, D. Scott, *Miracles* (Dial Press, New York, 1982).

136 Schatzman, Morton, *The story of Ruth* (Duckworth, 1980).

137 Schrenck-Notzing, Baron von, *Phenomena of materialisation* (Kegan Paul, 1920: first published in German, München, 1914).

138 Schwarz, Berthold E., 'Visiting with space people', in Curtis Fuller (ed.), *Proceedings of the first international UFO Congress, 1977* (Warner Books, New York, 1980).

139 Scornaux, Jacques, 'Et si Michel Monnerie n'avait pas tout à fait tort?' in *Lumières dans la nuit,* Nos. 177/178 (Chambon, 1978).

140 Ségard, Charles, 'Some reflections with regard to the phenomenon called materialisation', in *Annals of Psychic Science* (March 1906).

141 Sheldrake, Rupert, *A new science of life* (Blond and Briggs, 1981).

142 Shirley, Ralph, *The angel warriors at Mons* (Newspaper Publicity Co., 1915).

143 Shirley, Ralph, *The mystery of the human double* (Rider, circa 1930).

144 Slocum, Joshua, *Sailing alone around the world* (Sampson Low, circa 1900).

145 Smith, Robert D., *Comparative miracles* (Herder, St Louis, 1965).

146 Smythe, Frank, *Adventures of a mountaineer* (Dent, 1940).

147 Society for Psychical Research, *Journal:*
  a — Vol. 6, No. 104 — Lady B's case, 1893.
  b — Vol. 9, No. 172 — St Boswells case, 1900.
  c — Vol. 13, No. 240 — 'Dream romances', 1907.
  d — Vol. 17, No. 324 — Angels of Mons, 1915.
  e — Vol. 42, No. 717 — Landau case, 1963.
  f — Vol. 52, No. 793 — 'The Bournemouth poltergeist', 1983.

148 Society for Psychical Research, Proceedings:
  a — Vol. 4, part 8, Mrs Sidgwick, 'Notes on Phantoms of the dead', 1885.
  b — Vol. 4, part 8, ibid (Leigh Hunt case), 1885.
  c — Vol. 6, part 15, Myers, F. W. H., 'Recognised apparitions, ('scratched face' case), 1889.
  d — Vol. 7, part 19, Backman, 'Experiments in clairvoyance', 1891.
  e — Vol. 8, part 22, Morton, Miss, 'Record of a haunted house', (Cheltenham case), 1892.
  f — Vol. 10, part 26, 'Report on the Census of Hallucinations', (Kirk case), 1894.
  g — Vol. 12, part 30, Alice Johnson, 'A case of information supernormally acquired' (infected coat case), 1896.
  h — Vol. 18, part 47, Honeyman, John, 'On certain unusual psychological phenomena' (Mr A. case), 1904.
  i — Vol. 19, part 51, Piddington, J. G., 'Dr Henry Head's Goulstonian lectures for 1901', 1905.
  j — Vol. 27, part 69, Johnson, Alice, 'Pseudo-physical phenomena in the case of Mr Grünbaum, 1914.
  k — Vol. 33, part 86, Mrs Sidgwick, 'Phantasms of the living', (M'Connel case), 1922.
  l — Vol. 35, part 94, Leaning, F. E., 'An introductory study of hypnagogic phenomena', 1925.
  m — Vol. 44, part 151, Mace, C. A., 'Supernormal faculty and the structure of the mind', 1937.
  n — Vol. 45, part 160, Price, H. H., Presidential address: 'Haunting and the psychic ether hypothesis', 1939.
  o — Vol. 50, part 185, Hart, Hornell and others, 'Six theories about apparitions', 1956.

149 Spina, *Quaestio de strigibus* (Venezia, 1523), in Michael J. Harner (ed.), *Hallucinogens and shamanism* (Oxford University Press, New York, 1973).

150 Sprinkle, Leo, 'Hypnotic and psychic implications in the investigation of UFO reports', in Coral and Jim Lorenzen, *Encounters with UFO occupants* (Berkeley Medallion, New York, 1976).

151 Starbuck, Edwin Ditter, *The psychology of religion* (Walter Scott, 1901).

152 Steiger, Brad, *Mysteries of time and space* (Prentice Hall, New Jersey, 1974).

153 Stevenson, Ian, 'The contribution of apparitions to the evidence for survival', in *Journal of the American Society for Psychical Research*, Vol. 76, No. 4 (October 1982).

154 Stewart, Kilton, 'Dream theory in Malaya', in Charles Tart (ed.), *'Altered states of consciousness* (Wiley, New York, 1969).

155 Sudre, René, *Treatise on parapsychology* (Allen and Unwin, 1960: first published in French, 1956).

156 Symonds, John, *The magic of Aleister Crowley* (Muller, 1958).

157 Taylor, Gordon Rattray, *The natural history of the mind* (Secker and Warburg, 1979).

158 Thurston, Herbert J., *Beauraing and other apparitions* (Burns, Oates and Washbourne, 1934).

159 Thurston, Herbert J., *The physical phenomena of mysticism* (Burns, Oates, 1952).

160 Tizané, Emile, *Les apparitions de la Vierge* (Tchou, Paris, 1977).

161 Tizané, Emile, *L'hôte inconnu dans le crime sans cause* (Tchou, Paris, 1977).

162 Tyrrell, G. N. M., *Apparitions* (Duckworth, 1943: revised edition 1953).

163 Vyvyan, John, *A case against Jones* (James Clarke, 1966).

164 Webb, David, *1973, year of the humanoids* (CUFOS, Evanston, Illinois, 1976).

165 Wentz, W. Y. Evans, *The fairy faith in Celtic countries* (Henry Frowde, Oxford University Press, 1911).

166 West, Donald, *Eleven Lourdes miracles* (Duckworth, 1957).

167 Williamson, George Hunt, and McCoy, John, *UFOs confidential* (Essene Press, Corpus Christi, Texas, 1958).

168 Wilson, Beckles, 'The best attested ghost stories', in *Strand* (December 1908).

169 Wilson, Clifford, and Weldon, John, *Close encounters — a better explanation* (Master Books, San Diego, 1978).

170 Zurcher, Eric, *Les apparitions d'humanoïdes* (Alain Lefeuvre, Nice, 1979).

INDEX